The Low Carbon Diet

First published in 2007 by
Short Books
3A Exmouth House, Pine Street
London EC1R OJH

10 9 8 7 6 5 4 3 2 1

ISBN 978-1-904977-98-8

Illustrations by Tim Dowling and Dominic Cooper

Design by Two Associates

The **Low Carbon**
Diet

RECYCLE!

The Low-Carbon Diet is printed
by Cambrian Printers in Wales
on sustainable paper from certified
forests using vegetable-based
inks. When you have finished with
the book, please recycle it.

Contents

1

introduction

How to USE this Book ?**?**

(or The Low Carbon Diet and You)

This is a diet book with a difference. Yes, it can help you lose physical weight and save money. But its main purpose is to cut your "carbon calories," the everyday activities which make up your personal impact on climate change.

In Britain, as in all wealthy countries, we indulge in a carbon-heavy diet. Every time we flip a switch, turn on the car engine or buckle into a plane seat, we release gases that heat up the planet. Already, these are triggering extreme weather around the globe. If we carry on bingeing, the symptoms of our over-indulgence will get much, much worse.

Unless you never watch TV, read a paper, or have just returned from Outer Mongolia, you already know this. What we bet you don't know – but would like to – is how to adopt a low carbon diet. One that helps you to do your bit for a better future.

That's where this book comes in. Modelled on successful slimming guides, it's interactive, fun and definitely not preachy. **Here's how it works.**

Over the page is a to-do list of 20 Quick Fixes to stick on your fridge. This provides some simple effective actions to get started on right away. Next, there's a guide to calculating your personal carbon weight in tonnes of carbon dioxide, the main greenhouse gas. For those of you who want to take the easy route, we have made up some characters representative of British households and you can simply choose which is most like you and use their yearly carbon total. Otherwise, you can whip out a calculator and work out a precise figure based on energy bills and mileage totals.

The rest of the book explains the four golden rules of low carbon dieting and guides you through actions to take at home, in the garden and when travelling, eating, shopping or on holiday. As you start dieting, use our handy (and hopefully inspirational) Diet Masterplan. This lists 50 diet activities and the carbon calories saved by following each one, with boxes alongside in which to record your monthly progress over a year. And finally, once you've completed our standard Low Carbon Diet (we know you will), you may want to try out the carbon supersavers in the Advanced Diet which follows.

Good luck and happy dieting

A Word for Cynics The best guess of the world's top climate scientists is that we have 10-15 years to switch, en masse, to a low carbon diet. If we do that, most of the planet should remain habitable to humans (though probably not for polar bears and maybe not for penguins). If we don't, all bets are off. So forget about it being the government's job to fix this, or the bloke next door's, or the next generation. Just ask yourself: is it really too much effort to screw in a few energy efficient light-bulbs, insulate your loft or, next time round, buy a hybrid car?

Slimming Down:

20 Quick Fixes

Conventional wisdom says that to be a carbon dieter (or green consumer) you must spend more, give things up – or both. **No pain, no gain. This is a myth.** Here are **20** ways you can lose carbon weight for absolutely nothing. In most cases, by slimming down on energy use, you will come out quids in.

Garden

Don't use a patio heater or light bonfires.

Do compost your garden waste and/or use a water butt.

WELCOME HOME

Home

Turn down the heating by 1°C.
Sweater, anyone?

Avoid standby mode gluttony.
Turn off the TV, DVD player,
computer, stereo
(*you get the message*)
when you're not using them.

Find your water
heater and lower
the setting to 60°C.

**Flip the
light switch**
whenever
you leave
a room.

Put your central heating on a timer;
turn it off at night and wear warm PJs.

Unplug your mobile phone
as soon as it's charged.

Zap the tap when
brushing your teeth.

Hang out the washing on sunny days.

Re-use or **recycle**
paper, cardboard,
glass, cans, plastic
and clothing.

Fill the dishwasher and
washing machine
to the brim

Transport

Going a mile or less? Choose two legs over four wheels.

Keep your tyres inflated and the boot empty.

Don't drive to the gym, run there.

Lose that early am **stress**. Join a school run or a car share commute.

Shopping/Consumer

Buy wine from Europe, *not* Australia or California.

Do one weekly supermarket shop and buy in bulk.

Don't buy food flown into Britain; buy local as much as possible.

Holidays

Cut back on flights by at least one a year.

Where are you on the scales?

How do you rate on the carbon dieting scale? Are you pretty slim already or a real energy glutton? Give yourself one point for each activity above that you have already adopted. Then tot up your score.

15-20 Excellent

You're on your way to a really good diet. But remember these are only quick fixes. There's a lot more you can do by following the detailed advice in this book.

8-14 Not Bad

You're trying, but still pretty over-weight. There's so much more you could do, without much effort, to shed significant carbon weight. Read on!

1-7 Failing the Grade

Either you're too busy to make a bit of effort for the planet or you've slept through the whole global warming debate. Wake up and get dieting. **Please! It's not that hard...**

SO WHAT IS MY CARBON WEIGHT?

That little exercise gave you an idea of how much dieting lies ahead.
Now you need to estimate your total carbon weight so that you
can track how many kilos you're losing as you follow our diet.

Quick Take: Choose a Character

Take a look at the five households following and decide which one you or your
family most resemble (on a bad day). Then simply use that character's carbon total
as your own. This represents – roughly - your carbon weight as it is today.
When you start recording your diet's progress in our Diet Masterplan on page 135,
this figure will be the starting point from which you will deduct your weight
savings each month.

For example, say you decide that you most resemble
Danny. Your yearly carbon weight will be about 15 tonnes
or 15,000kg. If you save 300kg in your first month
of dieting, taking a few simple actions, your new total
will be 14,700kg or 14.7 tonnes.

Carbon weight totals for our characters are given per person. To record your
weight as a family, multiply the individual tonnage by the number of adults in
your household and add 50% of the same figure for each child.

Choosing a character will not give you a precise figure for your carbon consumption, but it will save you digging out a year's worth of heating bills and totting up your annual travel mileage. If you'd rather know your exact carbon consumption, then set aside a couple of hours, grab a calculator and go straight to **The Long Take on p.25**.

Character	Carbon Diet Rating	Carbon Weight (CO2 per person; cut by half for children)
Danny Chubb	**Obese**	15 tonnes
Diana Feather	Slim	5 tonnes
Brown Family	**Average**	10 tonnes
Watts Family	**Obese**	16 tonnes
The Greys	Slim	6 tonnes
My Household =	....	....

Danny Chubb

Danny Chubb lives life in the fast lane – literally. He loves fast cars, fast food and foreign travel. His pride and joy is a gas-guzzling BMW sports car in which he commutes from his Docklands flat to his City advertising job. He works long hours and rarely uses his designer kitchen, preferring to eat out or heat up readymade Thai and Indian meals, washed down with Californian wine. Energy efficiency is a foreign language to Danny. He uses his (council supplied) recycling bin to store his skis in and runs his home computer, TV and DVD player on permanent standby. He loves ultra hot power showers and in winter sleeps in boxer shorts with the heating on. Once a year he buys the latest mobile phone and chucks out the old one. Every few weekends he dons his shades and boards a budget flight to a European beach or city.

Diana Feather

Diana Feather is the anti-Danny. She works for a charity and is a fervent green consumer. She drives her friends crazy by rescuing recyclable silver foil and empty cans from their kitchen bins. She's "adopted" a penguin named Charlie in Antarctica and gives her family "plant a tree" certificates for Christmas.

Diana lives in a small flat and doesn't own a car. She takes the bus to work and cycles or walks everywhere else. She recycles everything, including old clothes, books and jewellery, has a super efficient boiler and long-lasting fluorescent light-bulbs in every room. She's re-plumbed her washing machine to run on rainwater and hangs clothes out on her patio to dry. After cutting out baths, she has decided to go one step further and only shower every other day, to promote a "natural body ecosystem." A vegetarian, she buys as much local and organic food as she can find and afford. Diana spurns flying and holidays in the UK. She loves yoga retreats and most recently joined friends in a communal "bunk barn" in Wales.

Watts Family

Charlotte and Charles Watts live in a five bedroom, detached suburban house, with a big, well-tended garden. Charlotte's favourite TV programme is *Desperate Housewives*; Charles loves the *Times* crossword. They have two cars, one an SUV. Both commute to work. The house is crammed with TVs, DVD players and computers – always on standby – to accommodate mum, dad, 12-year-old John and 14-year-old Fleur. The teenagers both have mobile phones and i-Pods. Fashion conscious Fleur demands (and usually gets) an entire new wardrobe every year. John's latest obsession is his mini motorbike.

This busy family eats a lot of convenience foods, especially burgers. In summer, they entertain guests around their outdoor barbeque and in winter around the patio heater. The heating, hot water, washing machine and lights are on 24/7 to suit everyone's different schedules. No time or thought is given to recycling – everything goes in the bin. Charlotte's prized possession is her orchids which she grows in a (heat and water intensive) greenhouse. Twice a year they fly abroad – to a European city to soak up some culture and to Florida. For her birthday, Charlotte treats herself to a long weekend in New York.

Brown Family

Tom and Linda Brown live in a three bedroom terraced house with daughter Sarah, 7 and 10-year-old Jack. They have a small, overgrown garden which they rarely water and a shed filled with bikes which only the children use. They own one car, a standard saloon, one TV and one home computer, which Jack is always colonising. Tom takes the train to work while Linda drives the kids to school. Once home, she walks to her part-time job at the local library. The family mostly eats and entertains at home. Linda buys organic milk and meat and eats her favourite fruits, pineapple and strawberries, year round. She can't stand being cold and Tom complains the house is like a sauna. They both try and remember to recycle, but half the time they forget the kerbside collection day and end up throwing glass and paper in the dustbin. Tom fancies himself as a DIY'er. He's draught-proofed the windows and doors and plans to insulate the draughty attic, but keeps putting it off. His weekend football games take precedence. The family flies to Spain or Greece most summers. Other time off is spent visiting family around the UK.

Nancy and Fred Grey

Both retired, Nancy and Fred Grey are energy and cost conscious. They live in a cosy two-bedroom cottage and use their small hatchback sparingly, for a weekly supermarket shop and social visits. Their biggest indulgence is driving 10 miles to the local town once a week to see a film or play. When they moved to their village, they insulated the attic and installed double glazing. Nancy is a bit of a self-sufficiency freak. She cooks and freezes in bulk, grows vegetables and herbs, makes her own jam and recycles food waste in her garden composter. She wanted to experiment with a beehive, but Fred managed to dissuade her. The Greys take barely warm showers ("good for the circulation", says Fred), and use their heating only in the depths of winter. Otherwise they make do with extra jumpers, hot water bottles and Tibetan wool slippers. Both compulsively switch off lights and tut at their grandchildren when they leave them on. They walk a lot for exercise and to local volunteer jobs, and holiday in the UK, by car or train.

Long Take: Calculate Your Own Carbon Weight

Filling in the chart on the next pages will give you a more accurate picture of your direct carbon consumption over one year than choosing a character.

Before you start, you will need to gather all your energy bills for the past 12 months. You will also need to calculate your annual personal travel, in kilometres, by car and plane; and estimate your travel in kilometres, by bus, train and underground. If you drive to work, include your daily commute in your car travel total, but don't include any business driving. You can calculate door to door distances in the UK on the Driving Directions page of www.viamichelin.com.

Hot Tips!

- For gas and electricity bills, you need the **total** consumption figure in Kwh – don't worry about different pricing bands. And it's OK if some of your bills are estimates.
- If your car is more than three years old, your two most recent annual MOT certificates will state the total number of miles you've driven in the car. To work out your annual mileage, take the figure in your most recent MOT certificate and deduct the mileage total on the previous certificate. (Example: year 1 = 9387 miles, year 2 = 20115 miles, annual mileage is 20115-9387 = 10728). If your car is new, divide the total mileage figure on the dashboard by the number of years (or part of a year) you've had the car.
- For travel on public transport, don't worry about getting totally accurate figures: just estimate the distance you travel to work (and back!). Don't forget to include an estimate of trips made for leisure, both long and short-distance.
- **Remember, to convert miles to kilometres multiply by 1.6.**

My Carbon Calculator

(Your personal yearly carbon footprint/kg CO2)

Home	My/Our Yearly Consumption (Divide totals by number of people in household to get personal total)	The Maths Bit (Multiply using a calculator)	My/Our Yearly Carbon Weight (kg CO2)
Electricity	kWh	X 0.43	....
Gas	kWh	X 0.19	....
Oil	Litres	X 3.00	....
Home Total			....
	Average UK home emissions per person:		2800kg

Travel (land)	(For car travel, divide by average number of people travelling to get the personal total)		
Car (petrol)	Km/yr	X 0.18	....
Car (diesel)	Km/yr	X 0.17	....
Urban Bus	Km/yr	X 012*	....

Travel (land) contd	My/Our Yearly Consumption	The Maths Bit	My/Our Yearly Carbon Weight
Underground	Km/yr	X 0.09*	....
Urban Train	Km/yr	X 0.14*	....
Coach	Km/yr	X 0.095	....
Intercity Train	Km/yr	X 0.04	....
Land Total			....
	Average person's travel emissions:		1600kg

Travel (air)			
Short-haul	Km/yr	X 0.405**	....
Long-haul	Km/yr	X 0.297**	....
Air Total			....
	Average person's travel emissions:		1971kg

Shopping

Which of the following descriptions best fits your lifestyle? Add up the figures to get total carbon weight (kg)

	Yes/No ✔ ✗	The Maths Bit	My/Our Yearly Carbon Weight
Car: I own a car	. . . .	+ 555kg	. . . .
Things: I buy second hand whenever possible	. . . .	+ 600kg	. . . .
or: I buy new things when I need them	. . . .	+ 2000kg	. . . .
or: I love shopping	. . . .	+ 3000kg	. . . .
Food: I eat meat	. . . .	+ 1200kg	. . . .
I only eat my home produced fruit and veg	. . . .	+ 0kg	. . . .
or: I mainly eat organic UK fruit and veg	. . . .	+ 400kg	. . . .
or: I mainly eat non-organic fruit and veg	. . . .	+ 800kg	. . . .
Shopping Total			. . . .

Average person's shopping total: 4000kg

The Low Carbon Diet

Your Total Carbon Footprint
(kg CO2) Add up the totals in each section.

. . . .

Average person's total carbon weight: 10, 371kg

Sources: DEFRA, Climate Outreach Information Network (COIN) and Dr Stephen Potter.

* for urban bus, underground and train, emission factors used are an average of peak and off-peak figures (CO2 emissions per person are higher during off-peak hours when there are fewer passengers, and lower during commuting hours when passengers numbers are high.

** emission factors have been multiplied by 2.7 to reflect the warming equivalent of other greenhouse gases in the upper atmosphere.

Now, whilst you're pondering your carbon weight (and perhaps getting cold feet), check out the climate change facts and images over the page.
They'll stiffen your resolve…

Why We All Need This Diet...

- Ten of the last 11 years have been the hottest on record. Up to 35,000 died as a result of Europe's 2003 heatwave.
- The culprit is climate change, which is bloating the planet's atmosphere and triggering extreme weather on every continent. 2005 was the worst year on record for climate related natural disasters.
- A carbon-heavy diet is feeding climate change. Every time we flick on a switch, turn the ignition or eat food flown across the world, our actions consume oil, coal, and gas and release carbon dioxide (CO_2), the main greenhouse gas. Ditto when we clear carbon-storing forests to grow crops or make paper, flooring and furniture. Methane (a nasty greenhouse gas 20 times more powerful than CO_2) is released through agriculture (yes, animals farting) and by rotting food waste.
- Every second, carbon-craving human beings shovel another 700 tonnes of CO_2 into the overloaded atmosphere – enough to fill 140 Olympic swimming pools.

- Every year the average UK citizen produces about 10 tonnes of CO_2 – enough to fill two Olympic swimming pools.
- If we do nothing, global average temperatures are forecast to rise by as much as 6.4°C by 2100. The last time we had a 5°C temperature swing (downwards), the world was locked into the Ice Age. As the world heats up, rising sea levels and stronger storms will swamp coastlines, extinguish species and could trigger the worst global depression since the 1930's.

To restore the Earth's health, those of us living in wealthy countries must dramatically restrict our carbon calories over the next 10 years.

The Good News (phew!)

You *can* make a difference **and** it will save you money.

Carbon dieting is **NOT THAT HARD**. Acting now to stabilise atmospheric CO2 would only cost around 1% of global GDP. For every pound spent now, say government economists, we would save a fiver in the future.

Individual dietary choices really add up. For example: transporting a single punnet (225g) of New Zealand strawberries to your fridge generates as much CO2 as 11 school runs. Leaving a 100W light bulb on for a mere half hour would fill a party balloon. Burning one gallon of petrol releases up to 10kg of the stuff. But walking or cycling is carbon-free.

There is only one solution to our CO2 weight problem: a low carbon diet. And the blueprint is right here in your hands.

Still not convinced?

Then turn the page

Present

These images document the devastating damage that climate change is already wreaking on our one and only planet. But what of the future?

By 2080, if climate scientists' predictions are correct, the world's maps will need to be redrawn. Expanding oceans and melting ice will send sea levels soaring, threatening coastal city dwellers across southern and Southeast Asia, swallowing low-lying Pacific islands and chunks of the Netherlands and flooding 30% of Africa's coastline. Scorching heat will wipe out not just the Emperor penguin and the polar bear but the Antarctic fur seal, tree frog and bowhead whale, and bring malaria to the doorstep of 290 million more people. Rainforests could disappear from large areas of Central Africa and Brazil, exacerbating the greenhouse effect. Deserts and diseases would spread and freshwater may become the planet's most precious commodity, with three billion more people than today enduring shortages.

Future

The Day After Tomorrow?

And what about life in the UK? Rises of up to 6.4°C over the next century – with a rise of 4°C probable – will mean wetter, warmer winters and drier, hotter summers. Children will grow up without knowing snow, but shrugging their shoulders at flooded city streets and emergency evacuations.

By 2025
(1°C average UK temperature rise):

? regular flash floods
? doubling of water bills to pay for new reservoirs
? malaria jabs for Mediterranean holidays

By 2050
(2-3°C average UK temperature rise):

? gardens disappear due to permanent hosepipe bans.
? London becomes a ghost city in summer as people flee the unbearable heat.

By 2080
(4-5°C average UK temperature rise):

? London is swamped by a storm surge as the Thames Barrier fails
? South coast beaches disappear as seas surge up to 2.5 feet above 2006 level
? Killer tropical diseases emerge from infested UK rivers

Thanks to: Pure – the Clean Planet Trust at www.puretrust.org.uk; An Inconvenient Truth, Al Gore, published by Rodale, 2006.

Of course, these predictions are based on a do-nothing scenario. If we act now, individually and as nations, this future need never happen.

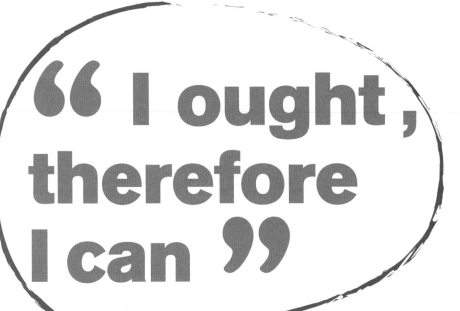

" I ought , therefore I can "

Immanuel Kant
German philosopher

2
My Diet
Building Blocks

Those are the stakes. So, now that we've got you on the edge of your seat and desperate to start shedding kilos, let's get started.

First – the good news. Cutting your carbon calories does not mean that you will a. be poorer or b. have to give up lots of things you love. These are myths put about by people who have a vested interest in you buying ever more stuff while paying no attention to the consequences for the planet. In fact, carbon dieting (like its older sister, green consumerism) simply means going about our daily lives in ways that consume less energy. And - more often than not - save us money into the bargain.

Fifty years ago, apart from the Queen and a few millionaires, we were all carbon slimmers. We lived in smaller houses, with fewer energy-hogging appliances, bought food produced on our own island or in Europe and took holidays in Britain. Far fewer of us owned cars and those who did drove smaller models, less often.

Today we live in a golden age of consumerism. We fly to Europe for £25, drive giant vehicles half as tall as our houses and buy strawberries in December, flown across the world to stock our 24/7 supermarkets. We even import Christmas. Last November, the world's biggest ship arrived in Felixstowe from China, laden with 45,000 tonnes of toys, clothes, crackers, and electronics for the festive season.

Are we suggesting a return to Fifties lifestyles? Of course not. This book is not about demanding the impossible. It's about using modern technology and science to consume less energy. It's also (let's be honest) about choosing a more moderate, back to basics consumerism driven by common sense and environmental awareness rather than debt-inducing one-upmanship and shopaholism.

Every single lifestyle decision we make – from what house we buy to how we get to work, where we holiday and what we eat for dinner – has carbon consequences.

By shopping smarter (hybrid car, anyone?) and living lighter, you can cut your carbon calories with minimal changes to your lifestyle. But negotiating the maze of modern consumerism can be confusing. So we suggest you use the following **Four Golden Rules** as a basic guide or framework.

Four Golden Rules

Save Energy and Money

Drive Smart, Fly Less

Buy Local

Slash the Trash

Save Energy and Money

Switch OFF and SAVE

The biggest diet-buster is home energy use. Heat, hot water and electricity account for 43% of the average family's direct carbon emissions, more than any other activity. But much of this energy is wasted, to the tune of millions of tonnes of CO2 a year. There is plenty of room to trim the fat, simply by cutting out careless habits and investing in a few home improvements. The key is to get into a mindset where it becomes second nature to save energy rather than squander it. After all, how hard is it to switch off a light?

Here's an added incentive: using less energy can also save you some dough. For example: £7 a year for each low energy light-bulb installed; £37 for switching off appliances on standby; £20 for draught-proofing windows and doors; £180 for insulating your attic. Not enough to retire on, but it will help the household budget.

Drive Smart, Fly Less

My next car's a hybrid – honest !!!

Why do you drive? OK, it's a provocative question, but think about it. To get to work despite the traffic jams, parking costs and road rage? To drive the kids half a mile to school, when you could all do with the exercise? To visit the local gym (duh!) or corner shop? Transport is the fastest growing source of UK greenhouse gas emissions. Before automatically grabbing your keys, think whether you really need the car this time. Again, it's all about mindset.

   We all love cheap flights. But here's the thing. Flying creates more personal carbon calories than any other single activity. Following our diet, then spending a long weekend in New York, is like losing a stone and then stuffing yourself on doughnuts and putting it all back on. We're not asking you to give up planes altogether. But finding other ways to travel, most of the time, is a cornerstone of a low carbon diet.

Buy Local

Easier said than done these days, given the global marketplace we all inhabit. But buying products flown thousands of miles to stock our supermarkets and high street chains, piles on our carbon weight. And if enough of us make an effort to buy local or UK made products, retailers will get the message that consumers don't want to shop at the planet's expense. So become a consumer detective and scrutinise labels for country of origin. Food, drink, clothes, furniture, toys, you name it. **Any product you buy, try and choose home grown or home made.**

Slash the Trash

Half a tonne of rubbish passes through the average British dustbin every year, enough to fill Wembley Stadium 52 times over. Most is buried in tips where the food waste decomposes and pumps out greenhouse gases. By throwing so much stuff away, we also create demand for new products, the manufacture, packaging and transport of which generate more greenhouse gases. Carbon dieters can turn this equation on its head by composting food waste, avoiding over-packaged goods and re-using or recycling their possessions.

How your Carbon Calories Add Up

To apply these rules most effectively, you need to know which of your daily activities have the most impact on the climate, so you can lose weight where it matters most. So far, we have talked mostly about your total carbon weight. From now on, we will be talking as much about "carbon calories" – i.e. the carbon energy you consume – and how to cut them.

As we said before, the average UK citizen's carbon weight is 10.4 tonnes. Of this, roughly half (5.4 tonnes, or the weight of an African elephant) is made up of carbon dioxide emissions produced directly by our daily personal activities – heating and lighting our homes, using appliances, driving our cars. The other half, described as "indirect emissions" are produced by the manufacture, production, processing and transport of the goods and services we use. These include the food and possessions we buy, the offices we work in and the hospitals, schools and other bits of infrastructure we use.

The carbon calories we have most control over are the direct emissions generated by our daily activities. The pie to the right shows how these emissions divide up. Of course, these reflect the lifestyle of the typical household. If you're one of the minority who don't own a car or fly on holiday, your pie will slice up a different way.

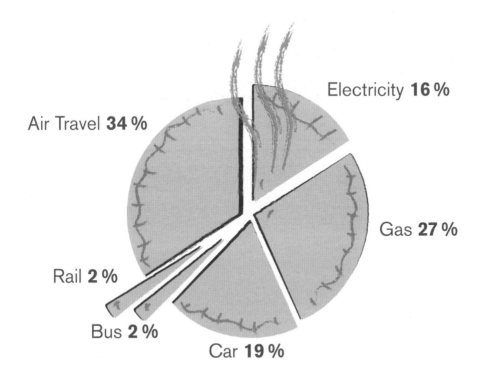

Air Travel **34 %**

Electricity **16 %**

Gas **27 %**

Rail **2 %**

Bus **2 %**

Car **19 %**

Which habits use the most carbon calories?

As you can see, home energy and car and air travel are the biggest culprits. On the bright side, as the next section describes, they are also relatively easy areas for big cuts. Specifically, the following bad habits are very carbon-intensive. Cut some or all of these out and you'll slim down in no time.

10 Worst
Diet-Busters

1 Owning an old (15 years +) leaky boiler

2 Not insulating your home

3 Over-heating your home

4 Never switching off –
lights, appliances, electronics,
you name it

5 Flying when you could
take the train or car/ferry

FLATTEN!

6

Driving a car which gets less than 30mpg

Owning more than one car

7

Shunning public transport

8

9

Shopping till you drop

Not giving a damn

10

What's the UK's carbon weight?

Now that the world has officially woken up to the danger of climate change, countries are carefully totting up their greenhouse gas emissions. Carbon dioxide, the biggest culprit, accounts for 84 per cent of these. In 2005, UK CO2 emissions officially totalled 561 million tonnes.

This ranks us among the world's carbon obese nations. Worse, after falling by 5.6% between 1990 and 2004, emissions are rising again and we will miss the government's much-touted target of reducing the UK greenhouse gas-o-meter to 20% below 1990 levels by 2010. The main culprit is road and air travel, but home energy use is also growing. Put simply, our love affair with motor cars, cheap flights, big houses and electronic gadgets is cancelling out the progress made by cleaning up dirty industries, introducing energy efficiency technologies and phasing out coal mines in favour of gas.

How nice it would be to blame all this on the government, the neighbours, or even our favourite punch bag, George W. Bush. But we can't. Yes, politicians could do a lot more to make low carbon living more attractive (big tax breaks on hybrid cars and free home insulation spring to mind). But individual behaviour multiplied by 28 million households is what is really driving this country's impact on climate change.

The better news is that our government is taking climate change more seriously than most. With sufficient political will and people power, by 2020 a slimmed down Britain could look something like this...

UNPLUG!

Low Carbon   Britain 2020

- Motorists use annual carbon rationing cards to buy petrol. If they bust their limit, they buy extra credit from people who don't drive or have fuel-efficient cars.

- Seven in ten new cars are hybrids, with sales boosted by generous tax breaks.

- A fifth of homes sport rooftop solar panels or mini windmills.

- Two-thirds of the food sold in supermarkets is UK grown.

How do we compare to other countries?

Surprise, surprise, the US tops the list of carbon gluttons. With 5% of the world's population, it produces 21% of global greenhouse gas emissions – 20 tonnes a year for every person. Next heaviest, per capita, are the Australians (very hot climate) and Norwegians (very cold climate). Behind them are Western Europeans, Canadians, Russians, Japanese and Middle Easterners.

The UK ranks 9th among the world's 194 nations.

At the other end of the carbon weight scales, 140 countries contribute one tenth of the global total. Most are in sub-Saharan Africa, which averages only 0.7 tonnes of CO_2 per person. To reach a decent standard of living, they need to consume more carbon calories (through universal access to electricity, for example), while other regions slim down.

Among wealthier nations, Sweden is a bright spot in bucking the trend towards carbon gluttony. Half its electricity comes from renewable resources, including water (hydro-electricity) and forest wastes, and buildings have been super-insulated since the 1930s. Got your notebook out, Gordon?

Want to know more?

National Energy Foundation
www.nef.org.uk/energyadvice/co2emissions.htm
www.defra.gov.uk/environment/statistics/

What works for you

Designing Your Low Carbon Diet

That's the big picture. Now back to you. The next step is to decide what lifestyle areas to concentrate on and, if you want to, to set a target weight loss figure for a month or a year. Every little helps and there is no set goal that is right for everyone. The key is to start slow and set realistic goals.

Take another look at our **Four Golden Rules**. Which would be easiest for you or your family to apply? For instance if you live in the country, need to drive a lot and can't imagine life without flying to exotic places, focus on energy savings in the home. If you rent a 10th-floor flat and can't control much of the energy loss from your building, but you live near a city centre, focus on changing your travel habits. Perhaps you could take the bus to work instead of the car and walk or cycle to the gym, supermarket and local shops. Slash the Trash and Buy Local should provide some carbon savings opportunities for everyone. *They just take a bit of time and effort to get started.*

When doing this, make sure that everyone under your roof is involved. You don't want rows with your partner because you're insisting he stops driving to work. Or mutiny among the kids because you've decided the TVs in their rooms must go. Changing ingrained habits takes time. If you've only got yourself to worry about, that's great. Be as radical as you like (in fact the more radical the better). But don't start lecturing your family about changing the world until their eyes glaze over.

Try explaining to the kids how simple actions like switching off lights save you money. Then offer to share that money with them at the end of the month, if they take part. Win over your partner by emphasising the challenge or the fun stuff (showering together, enjoying candlelit dinners?)

Set a Weight Loss Target

To keep yourself on track, you may want to set a monthly or annual slimming target. For example, say you've identified yourself with the Watts family. You could aim to cut your total, per family member, from 16 tonnes to 12 over a year. Or if our carbon calculator revealed that you're just a little overweight (say 11.5 tonnes) you could aim for a 10% cut. Here are some examples of personal carbon targets you could aim for over a year. To get a monthly target, just divide by twelve:

Take Your Pick: Personal Carbon Targets

UK average CO2 total in 2005	10.4 tonnes (per capita)
UK government target for 2010	8.32 tonnes (")
UK government target for 2050	4.16 tonnes (")
World average CO2 consumption	3.8 tonnes (")

Thanks to: World Bank Human Development Report 2003.

Hot Tip!

If you like the last two, you will probably need to go on our Advanced Diet.

Choose your dieting activities

Once you've decided what lifestyle areas to tackle, read the next section, The Low Carbon Diet. Have a piece of paper handy as you go along and jot down the most achievable actions for your circumstances. There are dozens to choose from in our Home, Garden, Transport, Consumer and Holiday diets.

Sample Diet

To help you visualise the real difference you'll be making through your carbon cutting activities, here's a sample diet. Let's pretend you're a real energy guzzler whose weight in CO2 is 50% above the national average. You'll be surprised – and, we hope, motivated – to see how much you can save without dramatic lifestyle changes. All the actions and savings listed are charted in our Diet Masterplan later on.

My Simple Slimming Plan
(Carbon Total: 15 tonnes)

Diet Action	Weight (CO2/kg) loss in 1 Month
Try out an energy saving bulb in the living room	3kg
Turn TV, DVD player and stereo off standby at night	13kg
Unplug computer at nights and weekends	12kg
Do a wash once a week instead of twice	2kg
Fill kettle with one cup of water	4kg
Turn heating down 1°C	25kg
Switch to French from New World wine (and love it!)	68gm
Jog to the gym (twice a week)	3.5kg
Walk to the deli for lunch instead of driving	7kg
Take Eurostar to Paris for a long weekend instead of flying	222kg
Total monthly savings:	292kg
Yearly savings (with four trips on Eurostar):	1723kg

Now let's fast forward a few months. Say you've really caught the dieting bug. Your gas-guzzling car starts playing up and you replace it with a super-efficient diesel Peugeot 107. You start recycling religiously and taking public transport to work. After one year of combining your Simple Slimming efforts and Crash Diet, your personal carbon calorie count has fallen by an impressive 7,099kg or 7 tonnes, to just under half your previous total. You now weigh in well below the national average. **Congratulations!**

My Crash Diet

Diet Action	Weight (CO2/kg) loss in 1 Month
Recycling	35kg
Commuting by Underground	58kg
Driving a Peugeot 107	95kg
Insulating loft	125kg
Two short haul flights instead of six	1,620kg (over 12months)
Total saving over 12 months:	5,376kg

Not sure you're still up for this? Or how much difference you alone can make?
Try pondering these wise words from anthropologist **Margaret Mead**,
over a local brew or a nice glass of French wine…

66 **Never** doubt
that a small
group of
thoughtful
people can
change the
world. Indeed,
it is the only
thing that
ever has 99

The Low Carbon Diet

Let's start making changes, beginning with your home and garden, then everyday travel, food, shopping and holidays.

Over the next 78 pages we provide detailed advice on how to trim carbon calories in every area of your life. The rewards, in both karma and cash, can be substantial. And the more actions you take, the more you save. Throughout the text we've used a large green **C** to highlight carbon savings and a **£** sign, where appropriate, for cash savings. We've also flagged up which of our four golden rules you'll be following by taking different actions.

At the end of the section is your Diet Masterplan listing 50 key actions you can take to reduce your climate impact. Each is accompanied by the monthly and yearly savings in carbon kilogrammes that you would make by taking that action. As you follow our diet, fill in the chart's calendar to track your total monthly and yearly weight loss.

Don't worry if you can't do everything or even most of the things we suggest. Carbon dieting is addictive once you get into the swing of it, but it takes time for the habit to become ingrained. And all of us have energy-wasteful pastimes (fast cars, holidays in Thailand, eating kiwi fruit) that we can't imagine giving up. **Just remember, every little helps.**

Good luck!

Home
Diet

"An Englishman's home is his castle" goes the old saying. Well, in 2006, many of our houses aren't much less draughty than if they still sported turrets and leaded windows. Of our direct carbon footprint, 43% is produced through heating and lighting our homes, more than any other activity. Three-fifths of this is gobbled up by space heating, one fifth by water heating and one fifth by cooking, lights and appliances.

The Home: An Energy Black Hole

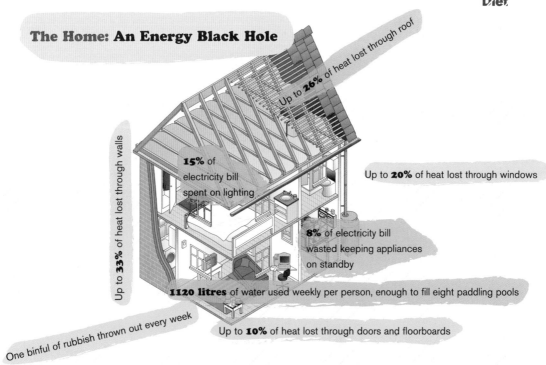

Up to **26%** of heat lost through roof

Up to **33%** of heat lost through walls

15% of electricity bill spent on lighting

Up to **20%** of heat lost through windows

8% of electricity bill wasted keeping appliances on standby

1120 litres of water used weekly per person, enough to fill eight paddling pools

One binful of rubbish thrown out every week

Up to **10%** of heat lost through doors and floorboards

Now the good news. Unless you already live an energy frugal lifestyle, the potential savings lurking under your roof are enormous. With a little effort and DIY muscle, the typical British family could slash significant carbon calories from their home. Our room-by-room tour will show you how.

First, though, a tip. Some changes we suggest require upfront investment and, before deciding on these, we recommend that you check out any local grants for home energy efficiency measures. Ring your local Energy Advice Centre on 0800 512 012 for free and independent advice. Or visit the Energy Saving Trust's comprehensive website at www.est.org.uk/myhome, click on 'Energy Saving Offers' and key in your postcode. If you decide to invest in new appliances, The Energy Saving Trust also lists products which carry an 'energy saving recommended' logo.

FYI: on average, we use 70% more electricity in our appliance-filled homes than our parents did in 1970.

Kitchen

Refresh the fridge and freezer

Freezer resemble a winter wonderland? Time to defrost.
Keep coils dust free. Dirt increases energy use by up to 30%.
Cool and cover foods before storing.
Replace damaged door seals – they let the heat in.
Fridge on its last legs? Replace with an 'A+' or 'A++'
rated appliance. Look for the appliance's energy label
to cut through the sales patter.

C 180kg **£** £45 *a year if you replace your fridge-freezer
with an A+ or A++ rated one.*

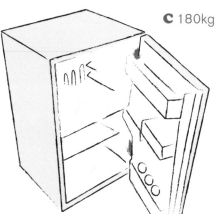

Rest the tumble dryer

Instead of that expensive Pilates class,
try stretching your arms while hanging
out the washing.

C 156kg **£** 37 *a year if you run the dryer half as often*
C 311kg **£** 74 *a year if you mothball it*

Fine tune the washing machine

You may be convinced that your whites come out best on a 90°C wash, but modern detergents are extremely effective. 40°C is fine, even stone cold, for those lightly stained items. While you're at it, experiment with cutting your weekly household washes by half, only wash when things really need it, and fit more in to each wash.

C 71kg £ 17 *a year if you lower the temperature to 40°C*
C 101kg £ 24 *a year if you run at 40°C and halve the number of washes*

Work the dishwasher harder

A 55°C cycle uses about a third less energy than a 65°C cycle. So pack in the dishes and set that economy wash button. It won't bite…

C 48kg £ 12 *a year for 5 cycles a week on an Eco setting*
C 101kg £ 24 *a year if you cut dishwasher use by half and use Eco setting*

Silent sinks

Don't wash up under a running tap – put a plug in it.
Deal with dripping taps. Believe it or not, each one can leak away
up to 140 litres a week.

Let the microwave sleep

Stick a reminder note on the microwave to switch it off at night.
Powering the clock alone can use as much energy in a year as is
used for cooking. Unless you're an insomniac, you won't need it.

Conserve that cuppa

Don't go over (or up to) the top.
Fill your kettle with the water you
need now, not the amount you might
need in four hours' time if someone
drops round.

C 48kg **£** 11 *a year*

Carbon-lite cooking

Some tips that Jamie isn't giving you. If you're boiling water, put a lid on it.
Uncovered pans lose 90% of their heat. Don't preheat the oven too early.
If you're defrosting food, or warming up leftovers choose the microwave.
It uses much less electricity and therefore fewer carbon calories
than conventional ovens.

Bin the old boiler

If your boiler is more than 15 years old, bin it. By law you must replace it with an 'A' or 'B' rated appliance, which means a high efficiency condensing boiler. This little dynamo will reduce heating-related CO_2 emissions by 15-20%. Even if your boiler's still young and attractive, consider trading it in. You'll shell out upfront, but long term you'll save.

C 1 tonne (1,000kg) *a year*
£ 100 *a year off your gas bill after paying back the purchase cost*

FYI:

If every UK home had a condensing boiler, CO_2 emissions would fall by 17.5 million tonnes a year and household energy bills by £1.3 billion.

Bypass the bin

Imagine not being able to chuck anything out for a week and watching the food and packaging waste piling up on your nice kitchen floor. Now hold that thought, start recycling all glass, paper, cardboard, cans and plastic and avoid buying over-packaged food that needs a bodybuilder to wrestle it open. If you have a garden, try composting all that food waste (Before you ask, there's a handy composting guide in our Garden Diet to come).

C 420kg *a year for recycling paper, plastics, cardboard, glass and cans*
C 280kg *a year for composting food/garden waste.*

Sitting Room

Lose standby mode

Train yourself to turn off those little lights twinkling on your hibernating TV, DVD player, computer or Hi Fi. Make yourself some **Switch Me Off** stickers to label different appliances as a reminder.

C 153kg **£** 37 *a year*

Unplug chargers

Another hi-tech solution.
Unplug appliance chargers when not in use.

Seal fireplaces

If you don't use it, board it up (try your hand at a mural). If you do, try this novel draught-prevention technique. Blow up a balloon, place it in the chimney, and have a good laugh when it bursts with that first winter fire.

Scary Statistics

7 in 10: the number of Britons who admitted to leaving electrical devices on standby in an Energy Savings Trust poll.

£255m: the amount of electricity a year consumed by DVDs and VCRs left on standby.

1 million: the amount of tonnes of CO_2 released annually by these DVDs and VCRs

Bedrooms

Total Turn Off

Just kidding. But we do want you to pile extra blankets on the bed, get out those PJs and turn the heating off at night. During the day, turn your thermostat down by 1°C and shed some serious carbon weight.

C 300kg **£** 50 *a year*

Curtains

Yet another hi-tech solution. Open them in the morning to let sunlight warm the room and close when it gets dark to keep in the heat.

Bathroom

Lose less from the loo

Loos are the biggest single water guzzler in the house. Fitting a Save-a-flush (a bag of harmless crystals) in your cistern costs only £1.20 and saves up to 1 litre a flush or nearly 2,000 litres per person a year (www.save-a-flush.co.uk). If yours is an old-fashioned inefficient loo, use a bigger Hippo (cost £1.32) and save up to 5,000 litres a year (www.hippo-the-watersave.co.uk).

C 0.58kg *for every 2000 litres saved; 1.45kg for every 5000 litres saved.*
£ *3% off your water bill with Save-a-flush, 9% off with a Hippo.*

Choose the shower

An average bath uses 80 litres (16 buckets) of water, a five-minute shower only 35 litres. It's also easier to share a shower than a bath (and save even more water) with someone you love.

C 200kg *if you fit a low-flow shower head (family of four)*

Brush teeth with taps off

Another of those bad habits we're all guilty of. Leaving the tap running uses six litres of water a minute. So run water only when you rinse.

Groom the old-fashioned way

Give up the electric shaver. And cut back on hairdryer use. Try the natural, windswept look. Or wait until your hair's almost dry before using.

FYI:

If all adults in England and Wales turned the tap off whilst brushing their teeth, it would save 180 million litres of water a day, enough to supply half a million homes.

Attic Give your loft a facelift

When was the last time you went up a ladder and peered into the unknown? It may be dirty and dusty up there, but insulating your leaky attic is a huge carbon calorie saver. Plus, it's cheap and easy enough to do yourself. The more insulation you put in, the less heat is lost. Building regulations for new homes require a minimum of 250mm (10in). Leave gaps around the eaves to avoid condensation.

C 1.5 tonnes (1500kg) *a year*
£ £180-220 *a year*

Basement, hall or landing

Coddle your hot water tank

Find out where it is and treat it to a nice, thick insulating jacket. Whilst you're there, check the water temperature setting – it should be no higher than 60°C.

C 150kg *a year when you fit a jacket on the tank;*
   145kg *a year when you turn the hot water tank down to 60°C.*
£ 20 *a year in reduced bills if you fit an insulating jacket.*

Get to know your timer switches

Most of us haven't a clue how to use them properly. We just turn the switch to **'on'**, forget about it and lose a lot of money in the process. Dig out your dusty instruction manual and set your timer to go off at night, when you're at work, or out for several hours. Most timers have a pre-set programme to help you out.

In Every Room

Seeing the light:

Replace your conventional bulbs with energy saving compact fluorescent lamps. Start in the entrance hall or landing – wherever lights are left on for a long time. They cost more than conventional bulbs (about £6 each) but use 75% less electricity and last eight times longer.

C 40kg a *year if you replace just one bulb*
£ 7 *a year per bulb replaced (savings on electricity bill minus cost of bulb)*

FYI:

90% of the energy burned in a conventional lightbulb is lost as heat. If every household installed three CFL bulbs, the energy saved in a year would power all street lighting in the UK.

FYI:

Tesco has pledged to halve the price of energy-efficient bulbs as part of a £500 million initiative to combat climate change. Check out your local store!

Keeping the heat in:

Draughts

This one involves a bit of detective work and a smoking stick – try incense or a cigarette. Carry this carefully around your house. Wherever smoke blows horizontally, you've spotted a draught. The most likely suspects are in fireplaces (at the base of chimneys), around window and door frames, and through letterboxes and cat flaps. Next, "draught-proof" your home by plugging up cold air sources with compression and wiper seals from a DIY store. Leave kitchen and bathroom windows alone to keep down condensation.

C 140kg *a year*
£ 20 *a year (less cost of seals in the first year)*

Radiators

This one has a touch of the Blue Peter about it. Simply slide aluminium foil behind any radiators fitted to outside walls to keep heat in the room. Ordinary kitchen foil will do, or you can buy specially designed panels from DIY stores. Also consider fitting thermostatic radiator valves (around £6 each) which control the heat level on individual radiators. Or do you really want radiators on continuously in unoccupied rooms, pumping out CO_2?

C 51kg **£** 7 *a year for foil*
C 110kg **£** 15 *a year for valves (after initial investment)*

Windows

If you don't have double-glazing, think about it. In both carbon and cost saving terms it's worth the hassle and money. And you'll be warm and toasty in winter. Start with the rooms that cost most to heat. Choose new windows with the 'energy saving recommended' logo. A budget alternative is to fit secondary glazing.

C 680kg **£** 90 *a year (after initial investment)*

Floors

The final word on draught-proofing, we promise. Fill those whistling gaps between old floorboards or the floor and the skirting board with commercial sealant. Or try papier mache. It could bring out the artist in you.

C 120kg **£** 10-20 *a year*

Garden
Diet

We love our gardens, so quintessentially English.
Who doesn't swoon over mown lawns, beautiful borders
and inviting garden benches? But here's the problem
for the carbon dieter. The things we put into our gardens
to make them green and lush are, directly or indirectly,
contributing to climate change.

Don't worry; you don't have to let your pride and joy turn into a wild,
weed-ridden mess. You can cut carbon calories with a few simple actions,
cheap investments and conservation-minded habits.
And your garden will still look lovely.

Calorie Cutting Do's and Don'ts

MADE
IN BRITAIN

DO discover drought-happy plants

When it's time to re-stock your flower beds, use plants that need less
water. Imagine your borders filled with beautiful grey lavenders, fragrant
rosemary and purple sage. These plants all originate in the Mediterranean,
are tough as old boots and very drought tolerant. Others that can withstand
long, dry periods and add interest to your garden are catmint, garden pinks,
French honeysuckle and rock roses.

DO water less

Fit your garden hose with a trigger spray system to cut down water use.
Better still buy a rainwater butt (see below) to cover your gardening needs.
If you have a large garden, consider installing a low volume irrigation system
with a timer, which can cut water use by half and the time you spend on tedious
watering duty by 90%.

To reduce evaporation, water early or at dusk, especially in summer. Instead
of frequent light sprinkling, water plants at their base thoroughly and infrequently.
A good trick is to bury the base of a plastic bottle upside down next to each plant,
directing water straight to the roots.

Use mulch to keep the soil cool and reduce evaporation. Plastic sheeting and
grass clippings work best. Or use home-grown compost.

Take a deep breath and let the lawn go a bit brown during summer. It can survive
long periods of dry weather if the grass is not too short. And with a couple of rain
showers it'll green up instantly.

FYI:
watering with an old-fashioned hand-held can uses four litres of water; a sprinkler system uses 540 litres an hour.

DO embrace organic gardening

Organic gardening is all the rage with Alan Titchmarsh (best-selling author of How To Be A Gardener), one of many celebrity converts. It's also a must for the serious carbon dieter. Organic methods shun manufactured fertilizers and other synthetic chemicals in growing plants and conditioning soil. By not using these products you will reduce your carbon impact since the energy used in their production releases greenhouse gases. You'll also encourage wildlife into your garden. And who doesn't love butterflies and bees?

DO propagate plants

The fewer plants you buy, the lighter your garden's carbon impact. Propagating is easy once you learn how. It's much cheaper to plant your own seedlings in old yoghurt pots than buy new annuals every spring. And it's carbon free – no driving to the garden centre. Cuttings also make great gifts.

DON'T use peat-based soil conditioners

Digging up peat bogs releases methane, a potent greenhouse gas. Peat bogs are also one of the UK's rarest habitats, yet 65% of commercial soil conditioners are still peat-based. Avoid like the plague.

DON'T light bonfires

They release CO2. Compost your garden waste instead.

DON'T use a patio heater

This directly releases CO2 through the fuel it burns. Wrap up warm and get a rug.

DON'T buy dodgy garden furniture.

Only choose items certified by the Forest Stewardship Council as coming from a well managed forest. Then sleep soundly knowing you're not sitting on a bench made from an ancient Amazonian tree. Of the major retailers, B&Q is the best place to find FSC-certified furniture. Or go to www.fsc-uk.org, key in the product you want and they'll tell you where to find it. Remember, buying local is best.

A New Take on Rainwater

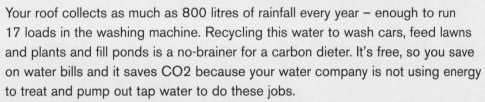

Your roof collects as much as 800 litres of rainfall every year – enough to run 17 loads in the washing machine. Recycling this water to wash cars, feed lawns and plants and fill ponds is a no-brainer for a carbon dieter. It's free, so you save on water bills and it saves CO2 because your water company is not using energy to treat and pump out tap water to do these jobs.

The easiest way to harvest rainwater is to place a water butt below your down pipe. Many water companies and local authorities offer special deals on water saving equipment. Make sure that you buy a complete water butt package which should include a diverter pipe to prevent overflow and a stand, so you can fill a watering can without breaking your back.

C 0.6kg *per year if you install a water butt*
£ up to £200 *off your yearly water bill*

Composting: the latest garden accessory

This is a great way to slash carbon calories. As much as 60% of household waste, including leftover food and paper, is biodegradable. Yet most goes to rubbish dumps where it decomposes and releases methane. Almost a quarter of the population now mucks in to compost kitchen and garden waste. So you'll be in good company.

Simply save food scraps from the kitchen and mix with garden clippings, fallen leaves and grass cuttings in a corner of the garden and hey presto! If you start in spring or summer, you'll have an excellent free soil fertiliser and conditioner in three to six months. Compost can also act as mulch, keeping those pesky weeds at bay. Your local authority may have a special offer on composters. Or make your own by cutting the bottom out of an old dustbin and propping it up on bricks.

C 280kg **£** 50

FYI:

Every year in the UK we landfill enough garden waste to fill the Royal Albert Hall more than 70 times over.

"My whole life has been spent waiting for an epiphany...the kind of transcendent, magical experience that lets you see your place in the big picture. And that is what I had with my first heap."
Actress and compost convert Bette Midler, quoted in the Los Angeles Times, May 1996

CARBON-LITE COMPOSTING RECIPE

PUT IN

FRUIT + VEG SCRAPS, TEA BAGS + COFFEE GROUNDS, CRUSHED EGG SHELLS, ANIMAL HAIR GRASS CUTTINGS, PRUNINGS + LEAVES SMALL AMOUNTS OF SHREDDED PAPER + SOFT CARDBOARD.

LEAVE OUT

MEAT + FISH SCRAPS, ANIMAL LITTER IF THEY'RE FED MEAT, ANY WASTE THAT CAN'T DEGRADE, WHOLE NEWSPAPERS OR UN-SHREDDED CARDBOARD, DISEASED PLANTS + WEED SEEDS.

No garden? Try worms…

Sounds dispiriting, but don't worry. We're talking about a sealed, hygienic, non-smelling unit, which can sit on your balcony, patio or kitchen windowsill. The worms live inside and you simply add kitchen and garden waste, paper and shredded cardboard. After a few months you'll have a ready source of natural plant fertiliser.

Transport Diet

Here's one simple reason so many of us are carbon obese: our cars.
We drive more and further than ever before, and buy bigger, thirstier vehicles.

Between 1990 and 2002, CO2 emissions from cars, planes and lorries rose
by a staggering 47%, says the National Office for Statistics. Imagine the impact
on your body if your food calorie intake grew by half. Then picture this.
If present driving habits continue traffic will rise by another 17% over 10 years,
and that's just in Britain.

On the **upside**, cutting the carbon emissions generated by cars isn't that hard
(no, really). We have the technology, the infrastructure, the information.
All we need is some collective dieters' willpower and a roadmap for change.

FYI:

we don't talk about flying in this section,
only road transport – but you're not off the
hook. Air miles are covered under the Holiday
Diet on page 128.

Your Car's Carbon Footprint

In a typical family, car travel accounts for one tonne of their direct CO_2 emissions, a fifth of the total. After home energy, it's the biggest cause of personal carbon calories.

This average, though, masks a wide range of consumer sins, based on the size of their car, fuel type, distances travelled and driving habits. A 4 x 4 driver who powers his way to 18,500km (12,000 miles) a year will produce emissions triple those of a diesel Vauxhall Corsa owner who covers the same distance. In other words, unless you already drive rarely and own a hybrid, there will be plenty of room to cut back your transport-related calories.

First, you need an accurate idea of your car's annual CO_2 emissions. You can work this out in two ways. Either use the sum from your Personal Carbon Calculator, reproduced below, which gives figures for an average sized petrol or diesel car in the UK.

Car:

Petrol (as driver) Km a year .... x 0.18 = ... kg CO2

Diesel (as driver) Km a year .... x 0.17 = ... kg CO2

Or, for a super-accurate result, spend five minutes on the Vehicle Certification Agency website at www.vcacarfueldata.org.uk. Enter your vehicle's make, model, fuel type and transmission and it will give you specific CO_2 per km figures for models going back to 1997. Multiply that figure by your annual kilometers travelled (a rough estimate will do) and you have your road transport carbon footprint. Alternatively, look up the mileage total on your last but one MOT certificate and deduct this figure from the total on your most recent MOT certificate. Remember, if you think in miles, 1.6km = 1mile. And if you're counting carbon emissions for the whole family and have two cars, don't forget to measure and combine the CO_2/km figure for both.

You may be pleasantly surprised by your results; or horrified. Older cars often have higher emissions than some bigger, newer makes because of the strides made in fuel efficiency in recent years. But if you've been seduced by a gas-guzzling 4 x 4 or a high-octane sports car your emissions (and motoring costs) will be very high. On the other hand, the new breed of small, energy-light smart cars yield both high mileage and low CO_2 emissions. If you drive one of these, you're already well ahead of the pack.

Wherever you fit in, the next step is to cut your down your monthly total. Start with smarter driving and cutting back car miles for short journeys and commutes. If you need to change cars, consider an energy-efficient model. You'll be amazed how much money you'll save.

Carbon Smart Driving Tips

We all do it. Rush out the door late; screech up the street; accelerate away from traffic lights like Nikki Lauder; forget to service the car every year (it runs, what's the problem?); let our tyres go bald; race other drivers on the motorway.

None of these bad driving habits are good for either our car or our blood pressure. They also waste fuel, and release a surprising amount of carbon dioxide. If you do *all* the following, *all* the time you can slash 10% off the carbon you emit in a year's driving. None of them cost anything – except for getting a service and you need to do this anyway.

Plan Ahead: choose the least busy routes, combine trips, carry passengers.

Don't Dither: drive off as soon as possible after starting.

Think Smooth: avoid sharp braking or acceleration which guzzles up petrol.

Slow Down: you burn 2% more fuel for every mile per gallon over 55mph.

Move up a Gear: as soon as traffic allows, it uses less fuel.

Switch Off: if you're going nowhere for at least two minutes.

Lose Weight: empty the boot, remove unused roof racks.

Service Often: for a super-efficient engine and a smooth drive.

Avert Tyre Trauma: keep at the correct pressure and save 5% on your fuel bill.

Avoid the Air-Con: running air-conditioning increases your fuel use.

C *10% of your car's annual CO2 output.*
£ *85 a year in petrol bills for an average size petrol car.*

Trim Those Car Miles

A quarter of all journeys made in the UK are 3km (2 miles) or less, which makes them ideal journeys on which to leave the car at home. So are commutes, as long as you live in an area well serviced by public transport.

Before you start mumbling those lame excuses, think of the benefits. Walking or cycling instead of driving can help you: lower blood pressure; lose fat; cut the risk of heart disease; breathe fresher air, (yes, air quality is often worse inside cars than outside); and improve time management (– no jams, parking problems or broken meters to slow you down).

So here goes...

Target 1: the dreaded commute

If you work within 2km (1.2 miles) of your home, try getting up earlier and walking or cycling to the office. Before you know it, you'll be able to cancel that expensive gym membership. If you enjoy a lie-in or have a longer commute, try the bus, underground or train. You can catch up with work, make phone calls, write a novel, do the crossword puzzle, you name it. If you have flexible working hours, travel before or after the 7.30am-9am rush.

Use the following table to work out your weekly carbon calorie savings if you switch from driving to alternative modes of transport.

Multiply the CO2 figure for travel by train, bus or underground, by the number of kilometres travelled each month. Then calculate your savings by deducting this figure from the total you would have incurred if you had driven. Car emissions during rush hour, by the way, are much higher than the daily average – 339g/km rather than 180g/km for an average size petrol car – so you'll be shedding significant calories by taking the bus instead. And if you choose the healthiest options – walking or cycling – you'll also pocket cash from all those gallons left un-pumped.

The Carbon Commute

	CO_2 emissions*
By diesel train	98g for 1km
By Underground	65g for 1km
By single decker bus	66g for 1km
By foot/bicycle	0g for 1km
By average petrol car (rush hour)	339g for 1km

*emissions during peak commuting hours

(Thanks to: Dr Stephen Potter, Open University; full reference on page 210 = Assumptions page)

If you're not sure how long your journey is and you live in London, visit www.walkit.com which will work out the distance for you. Outside the capital, use the AA's handy website at www.theaa.com/travelwatch/planner_main.jsp to get distances between destinations nationwide.

Keep up your resolve by recording your monthly savings on the Diet Masterplan. Here are some sample calorie savings if you leave your car in the garage:

C For a daily 12km round trip commute: train = 51kg; bus = 57kg; tube = 58kg.

C For a 4km roundtrip walk to work: 24kg

C For a 12km daily round trip by bicycle: 71kg

Target 2: the crazy school run

Easy one this. At 8.50am almost one in five cars in UK towns and cities are on a school run, creating a giant mushroom cloud of carbon dioxide. If your kids' school is 2km (1.2 miles) away or less, use your legs or bike. Research shows that more active children are likely to become more active, healthier adults. So get up 20 minutes earlier, prepare the kids' lunch the night before, do whatever it takes to get out the door with enough time before school starts. If ditching the car every day is too daunting, try a two or three day a week commitment. That way you have a get-out if it rains. Sample savings:

C 26kg *a month for a 4km round trip, twice a day*
(if you bike/walk to school 3 days out of 5).

Target 3: other short hops

These include trips to the supermarket, the local shops, the library, friends' homes, your health club. If you find walking dull, make it interesting. Put on headphones and a CD or iPod. If you need to work, take a tape recorder and dictate as you walk. If you're carrying things home, take a backpack or one of those trendy new shopping trolleys, or use a bike with a pannier. Sample saving:

☾ 11.5kg *for 20 short trips a month out of rush hour (64km total travel)*

LA LA LA
LA LA LA
LA LA
LA LA

Longer Journeys: Car, Train or Coach?

For longer commutes and work journeys it makes even more sense to let a train or coach take the strain. Think less stress, road rage, traffic jams, excuses for late meetings etc. At first glance the petrol costs may come in cheaper than the train season ticket, but then there's all the wear and tear on your car to factor in.

Use our Carbon Commute table on page 95 to work out your CO2 savings per roundtrip. Again, if you're hazy about the distance involved, get help from www.theaa.com/travelwatch/planner_main.jsp. If your regular journey is during peak hours, use 339g/km to calculate your car emissions per trip, if it's off-peak, use 180g/km.

Not convinced? Check out these carbon weight savings on popular long distance UK commuter routes. Then give it a go.

Biggest Carbon Footprint

Want to know more?

www.transport2000.org.uk
www.eta.co.uk

Medium Carbon Footprint

Smallest Carbon Footprint

Where To?	Distance (roundtrip in km)	Carbon Emissions (CO2) Car	Carbon Emissions (CO2) Coach	Carbon Emissions (CO2) Train
London to Reading	130	23kg	12kg	5kg
Edinburgh to Glasgow	144	26kg	13kg	6kg
London to Oxford	198	36kg	18kg	8kg
London to Brighton	168	30kg	16kg	7kg
London to Winchester	218	39kg	20kg	9kg
London to Bristol	380	68kg	35kg	15kg
London to Northampton	212	38kg	20kg	8kg
London to Manchester	638	115kg	59kg	26kg
London to Cambridge	198	36kg	18kg	8kg
London to Southampton	256	46kg	24kg	10kg

Using DEFRA figures for car and train, DfT figures for coach

The next car you buy could really slash your carbon calories.

The difference in C02 emissions released by the most fuel-efficient (non hybrid) car on Britain's roads, the Toyota Aygo, and the biggest gas guzzler, the Lamborghini Diablo, is 411 grams per kilometre. That's 2.8 tonnes a year, based on the average annual car use of 16,000km (10,000 miles)! OK, not many of us drive Lamborghinis. But the carbon impact of popular models also varies enormously. The diesel version of Britain's top-selling car, the Ford Focus, for example, releases a respectable 129g/km of carbon dioxide while a Ford Mondeo estate spews out 226g/km.

Of course you don't just want a climate-friendly car. You also want to drive in style and comfort and have plenty of room for the kids or your skis and surfboard. **No worries.**

Just digest our buyers' tips, then review the 10 Best Petrol and Diesel Cars we highlight below and try test-driving these models. Hopefully you'll find one you like that fits your needs and budget. We've also highlighted good carbon-lite buys among popular cars. Hybrids are not in this section – as they're still a small, expensive market. But you can find out all about them in our Advanced Transport Diet on page 170.

FYI:

automatic cars can burn 10-15% more fuel than manual models.

Buyers' Tips

Due to strides in fuel-efficient technologies, new cars are better than older ones (unless you're swapping a Mini for an SUV). Nearly new is even better, because you won't be responsible for the energy used in manufacturing a new vehicle.

Diesel is more fuel-efficient than petrol, produces less CO_2 per km and is therefore a good bet for the carbon dieter. However, it does produce nasty particulates (or soot particles) which cause air pollution. The answer is to buy a vehicle with a particulate filter, or get one fitted. They cost a few hundred pounds, but you'll save that on fuel bills over a year or two if you're a petrol convert.

As our Top 10 table below shows, small super-fuel-efficient cars can save you as many carbon calories as pricier hybrids, such as the Honda Civic. This is because miles per gallon equate directly to the greenhouse gases released by car exhausts.

By law, labels giving the A-G fuel efficiency ratings of new models must be displayed on all new showroom vehicles. These are based on CO_2 emissions per km. A carbon dieter should choose from grades A to C.

New vehicle excise duty rates for petrol and diesel cars were introduced in 2006. Based on CO_2 emissions, tax discs now range from only £30 for hybrids (hurray!) to £215 for Band G gas guzzlers. Trade your SUV for a Band C family car and you would save £105 a year. To find the VED band of cars you're interested in, go to www.vcacarfueldata.org.uk/ved/index.asp.

What Car?

Take your pick from the Top 10 carbon-lite petrol and diesel cars below, as ranked by the Vehicle Certification Agency. The list includes more mainstream diesel models (Ford Fiesta, Renault Clio) than petrol ones, due to their extra fuel efficiency. However, the Vauxhall Corsa, one of Britain's ten best-selling cars, makes it onto both lists. The lists are based on test drives of new models by the VCA (in other words, they're kosher).

Top 10 Carbon-Lite Petrol Cars

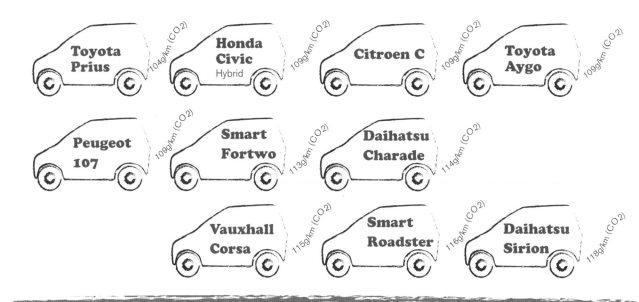

Toyota Prius — 104g/km (CO2)

Honda Civic Hybrid — 109g/km (CO2)

Citroen C — 109g/km (CO2)

Toyota Aygo — 109g/km (CO2)

Peugeot 107 — 109g/km (CO2)

Smart Fortwo — 113g/km (CO2)

Daihatsu Charade — 114g/km (CO2)

Vauxhall Corsa — 115g/km (CO2)

Smart Roadster — 116g/km (CO2)

Daihatsu Sirion — 118g/km (CO2)

Top 10 Carbon-Lite Diesel Cars

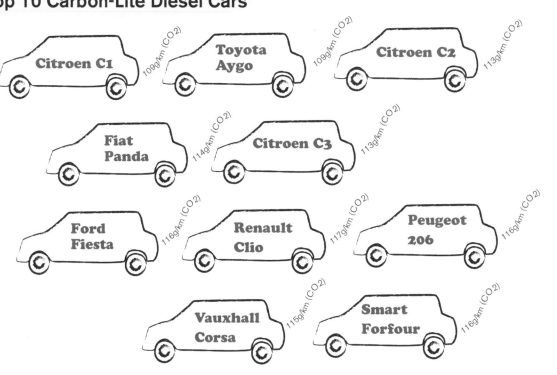

Citroen C1 — 109g/km (CO2)

Toyota Aygo — 109g/km (CO2)

Citroen C2 — 113g/km (CO2)

Fiat Panda — 114g/km (CO2)

Citroen C3 — 113g/km (CO2)

Ford Fiesta — 116g/km (CO2)

Renault Clio — 117g/km (CO2)

Peugeot 206 — 116g/km (CO2)

Vauxhall Corsa — 115g/km (CO2)

Smart Forfour — 116g/km (CO2)

10 Worst Carbon Hogs

Make	Model	CO2 (g/km)	Fuel Costs (12000 miles)
Lamborghini	Diablo	520	£3357
Bentley	Arnage RL	495	£3186
Aston Martin	Vanquish	448	£2929
Bentley	Flying Spur	423	£2728
Aston Martin	206 Coupé	421	£2762
Rolls Royce	Phantom	385	£2758
Volkswagen	Phaeton	374	£2411
Chrysler	Jeep New Grand Cherokee	366	£2385
Corvette	C6	350	£2273
Porsche	Cayenne	317	£2039

Thanks to: Environmental Transport Association www.eta.co.uk

Best-Selling Car Comparisons

If none of the top 10 models catch your fancy, be aware that many best-selling cars in the same size and price brackets vary widely in their fuel efficiency. Here are some examples of the cash and carbon savings to be made by switching between popular UK models, courtesy of the VCA website. Miles per gallon figures are for city and highway driving combined.

Large family car: Exchange a Land Rover Freelander (22.7mpg) for a Ford Focus (59mpg).
Save: C 119kg every 1000km
£ 927 every 12,000 miles (18,000km)

Small family car: Trade in a Volkswagen Golf (35mpg) for a diesel Ford Fiesta (64mpg).
Save: C 71kg every 1000km
£ 592 every 12000 miles (18,000km)

Small car: Trade a Vauxhall Astra (40mpg) for a diesel-powered Vauxhall Corsa (63mpg).
Save: C 50kg every 1000km
£ 436 every 12,000 miles (18,000km)

Want to know more?

www.vcacarfueldata.org.uk
www.whatcar.co.uk Click on Help
and Advice, then Going Green.

Feel-good Extras

Swap your AA membership for one with the *Environmental Transport Association*. True to its name, the ETA campaigns for climate-friendly transport policies. It offers full breakdown services for four-wheel vehicles as well as bicycle insurance. Go to www.eta.co.uk.

Join BP's *Target Neutral* scheme and "offset" (ie. cancel out) your car fuel emissions by paying toward renewable energy projects in developing countries. It costs around £20 for the average car, driven 10,000 miles a year. Sign up at www.targetneutral.com.

Choose *Co-operative Insurance's* (part of the Co-Op Bank) vehicle eco-insurance service. On offer are discounts on premiums for drivers of more climate-friendly (fuel-efficient) cars and a repairs service that recycles oil, bumpers and other materials.

FYI:

The Government is to introduce a new voluntary standard for carbon offsetting schemes, which aims to reassure consumers that their money is being spent on projects that genuinely reduce CO_2 emissions. The standards will be based only on projects that can be certified, including flexible schemes agreed under the Kyoto protocol on climate change.

Consumer Diet

Buying power is one of the greatest weapons available to the carbon dieter. Simply by eating and shopping we generate an average 3.5 tonnes of CO_2 emissions a year. Changing our buying habits can slash this total while improving your bank balance. But making carbon lite choices is not always easy, especially when retailers and manufacturers supply little or no information. In this section, we guide you through the shopping mall maze with advice on how to choose food, clothes, everyday essentials and big purchases such as TVs. Follow our simple rules and you won't go far wrong.

Food

Food production, processing, packaging and transport account for at least 20% of UK greenhouse gas emissions and a sixth of the carbon emissions produced by a typical household.

What action can you take? The answers are not all simple or obvious. Organic food is not always a good option, especially when delivered by plane. Food miles are bad, but long distance delicacies do not generate the most food-related emissions. That dubious prize goes to animal rearing, which means eating meat is bad for a low carbon diet. Other culprits include food processing, refrigeration and supermarket storage. To keep it simple, we suggest you follow these five golden food rules:

Carbon Dieter's Five Golden Food Rules

MADE IN BRITAIN

1: Buy Local and In Season

2: Eat Less Meat

3: Eat all the Food you Buy

4: Buy Organic from the UK

5: Buy in Bulk

With thanks to: Tara Garnett, coordinator, Food Climate Research Network.

1: Buy Local and In Season

Not so long ago, this was the norm. But in our new global food market, we fill our fridges with food and drink from a hundred different countries, 24/7, 365 days a year. The average large supermarket now stocks a mind-boggling 26,000 items.

Why does this matter? In a word, transport. Cars, lorries and planes cover 11bn km a year moving food around Britain, releasing 18 million tonnes of carbon dioxide, according to Defra. A growing percentage is flown long haul, leaving a huge CO_2 vapour trail in its wake.

The closer to home your food originates, the smaller its impact on the atmosphere and the more carbon calories you shed. Good choices, because they're usually UK grown, are carrots, parsnips, turnips, potatoes, cabbage, sprouts, broccoli, apples and pears. A majority of supermarket chicken, pork, beef and lamb is also home-grown, but always check the country of origin. The difference you can make is well worth tinkering with your diet. By buying British instead of foreign produce, you can save almost 54kg/CO_2 on the contents of a single shopping basket.

Made in Britain basket contains: cauliflower from Lincolnshire, mushrooms from Ireland, brussel sprouts from Lincolnshire, broccoli from Worcestershire, carrots from Scotland, onions from Shropshire.
Total = 0.18 kg CO2

Flown from Abroad basket contains: limes from Brazil, pears from Italy, avocados from Chile, peaches from USA, pineapple from Costa Rica, baby corn from Kenya.
Total = 54kg CO2

Where to Buy?

If polls are to be believed, three-quarters of us want to buy more local, in season food. But supermarkets don't always offer much help, forcing you to play detective up and down the aisles. So what's a carbon dieter to do?

One approach is to use a local farmer's market, farm shop or box delivery scheme as your main source of produce, assuming there's one nearby and you can live with the price mark-up. For those in your area, see www.bigbarn.co.uk, www.farmersmarkets.net or www.farmshopping.com.

When you do use the supermarket, be a pest. Ask for help finding UK sourced choices for everything on your shopping list. Bug the customer help desk for displays on in-season fruit and vegetables. This should soon become easier as supermarkets scramble to respond to public concern over food miles. In January 2007 both Marks & Spencer and Tesco announced plans to mark all products transported by air with an aeroplane symbol. Tesco also pledged to label every one of its 70,000 products with a carbon footprint while M&S committed to send zero waste to landfill by 2012.

Hot Tip!

For a top chef's advice on seasonal recipes, see Hugh Fearnley-Whittingstall's fun and informative website at www.rivercottage.net/

2: Eat Less Meat

This could be the single biggest difference your eating habits can make to your carbon diet. Rearing animals and processing and refrigerating meat products uses much more energy than growing crops, fruit and vegetables. Plus, livestock release vast quantities of methane through burping and farting. Cows are the worst offenders. Each 1.1 kg (1lb) of UK reared beef on your dinner table has cost 5.2kg of CO_2 to produce.

Enough to put you off your Sunday roast? Try reducing your intake of beef, steak, mince and burgers by a meal or two a week. Eat more beans, fish, fruit and vegetables. Diet gurus are urging a similar recipe to improve Britons' health and weight, so you win all round.

C 2.6kg *for every beef meal you cut out. Eat beef products twice a week instead of four times and save 21kg/CO2 a month.*

FYI: How fresh is that meat anyway? To reach UK tables, Australian beef goes through up to 24 processing and transport stages, reports the campaign group Sustain.

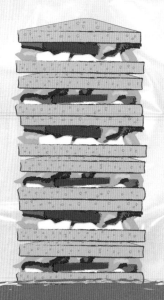

3: Eat All the Food you Buy

Sounds easy, but we typically waste a lot more food than we realize.
Over a year we throw out a fifth of the fresh produce we buy – £424 worth –
according to the Food Climate Research Network. If we ate all that fruit, vegetable
and salad, imagine the domino effect. Demand, domestic production and imports
would all fall and so would national food-related CO_2 emissions.

So do your bit. Plan your shopping carefully, meal by meal, don't buy
more perishables than you can eat in a week and pay close attention to those
use by dates.

4: Buy UK Organic Food

Unlike buying local, organic food is already a national habit. Three in four of us regularly buy organic produce, dairy products, meat, tea or coffee and so on. This is good news for the planet as less energy is generally used to produce organic foods than by conventional agriculture.

On the other hand, UK demand far outstrips supply which means most organic products are imported. And the transport involved almost always produces more CO_2 emissions than those saved through organic production – 235 times more in the case of organic produce flown from New Zealand. Even trucking fruit or vegetables from southern Europe generates twice the energy saved during organic production. The golden rule? Buy organic products made in Britain (fruit, vegetables, eggs and meat are all widely available). But choose local food over overseas organics, especially if imported by air.

Hot Tip!

don't choose local if it has been grown out of season, in a hothouse, like tomatoes or strawberries.

5: Buy in Bulk

We each drive an average 130 miles a year just to buy food. That's a lot of carbon calories when multiplied by the two-thirds of UK homes that own cars.

The simple solution is to buy in bulk. If you usually shop for food two or three times a week, aim for once. If you also visit a farmer's market, combine the trips to cut down on mileage. Non perishables – cereals, rice, pasta, juice, tinned goods, pet food, bottled water etc - are cheaper to buy in bulk and as a bonus you'll cut down on packaging waste. If you do run out of bread, milk or some other essential, walk to a local shop. Home delivery is another option, with the supermarket doing the work and saving food miles by delivering to several customers on one trip.

☾ Do one weekly supermarket shop instead of three – 4kg *a month*, 52kg *a year*

Diet-Busters to Avoid

Top 10 UK Fresh Fruit and Veg Air Imports	Annual tonnes of CO2 emitted by flying them to the UK
1. Green beans	88,000
2. Peas	46,000
3. Grapes	44,000
4. Mangoes and guavas	36,000
5. Sweetcorn	34,000
6. Pawpaws	33,000
7. Asparagus	26,000
8. Cherries	21,000
9. Pineapples	20,000
10. Onions	17,000

From Plough to Plate by Plane, by Clive Marriott, Food Climate Research Network, University of Surrey.

The Low Carbon Diet

The Carbon Conscious Shopper

It's hard to resist the advertising blitz and peer pressure that has turned us into a nation of shopaholics. But, boring as it may sound, the single most effective diet action you can take as a consumer is not to buy anything that you don't really, really need. So before you reach for the latest I-Pod, remember: anything new you buy will add to your carbon calorie count for the simple reason that it took energy to manufacture.

To help you curb your inner shopper, take a look at the rewards you could reap over a year:

C 1 tonne *if you only bought new things when you needed them rather than on a whim.*

1.2 tonnes *more if you also bought second hand rather than new whenever possible.*

When you do buy something new (and let's face it we all do, pretty often) use our five simple rules as a guideline. They will help you find the most carbon-friendly products and reward retailers that are making an effort to limit their carbon footprints.

Carbon Dieter's Five Golden Shopping Rules

1: Check Labels

2: Buy Local

3: Avoid Disposable Products

4: Choose Recycled Content and Recyclable Parts

5: Avoid Heavy Packaging

1: Check Labels

When we like the look of something and the price is right, we tend to just buy it. But for carbon dieters, labels and information leaflets carry a wealth of information to inform good purchasing habits. When you make any big purchase – from household appliances to furniture, curtains, electronics and electrical goods – look for answers to these questions. Where was the product made? What is its energy efficiency rating? Are any of its materials or parts recyclable? Does the manufacturer offer recycling take back schemes? Is there any environmental endorsement, such as the FSC logo on wood products or the EU's flower-shaped Ecolabel?

If information is lacking, talk to shop assistants or the store manager. For energy-guzzling products such as computers and TVs, check out consumer websites and magazines for the most efficient models. Which? magazine, for example, identified TVs made by Sony, Ferguson, Matsui, Samsung and Sharp as using the least energy on standby. Or visit www.gooshing.co.uk, the Good Shopping Guide's website, which compares thousands of products on environmental criteria including energy efficiency and organic content.

2: Buy Local

Three little words, Made in China, sum up the consumer revolution whereby most of what we buy these days was made abroad generating huge volumes of greenhouse gases in transit. Cheap prices are hard to resist. But, as with food, wherever reasonable your first choice should be to buy local, second to buy UK, third to buy goods made in Europe and lastly to buy from Asia or the Americas. Try visiting local high street boutiques, gift shops and craft outlets more likely to feature locally made products than chain stores. Online shopping for UK made items is another good bet.

3: Avoid Disposables

Say goodbye to plastic razors, paper towels (remember napkins?), non rechargeable batteries, throwaway cameras, plastic cups, plates and cutlery…you get the idea. You'll cut carbon calories because the high turnover of disposable products means more energy is generated making and disposing of them than non-disposable alternatives.

4

4: Choose Recycled Content and Recyclable Parts

Before you recoil, recycled content is not the same as second hand. These days, many brand new products are made using some recycled parts or materials – wood, paper and plastic being the most obvious. Paper products, such as toilet rolls, paper towels and envelopes generally display their recycled or "post consumer waste" content. Clothes are also well signposted. With bigger items such as electronics or household appliances, you may need to browse through operating manuals or manufacturers' websites.

The other side of the coin is to look for products with recyclable parts that you can return to manufacturers. Ecover, for example, offers a bottle refilling service for its household cleaning products. Most mobile phone and some computer companies operate take back schemes. All the major battery manufacturers will recycle their used products.

5: Avoid Heavy Packaging

So ubiquitous have layers of plastic wrapping and cardboard boxes become that 40% of the rubbish in a typical UK dustbin is retail packaging. To save carbon calories (and protect your fingernails), avoid any product with more than one layer. Refuse carrier bags when buying small items that can go straight in your handbag or rucksack. In supermarkets, buy loose fresh produce rather than the shrink-wrapped variety. Where possible, choose items packaged in cardboard rather than plastic, which is accepted by fewer recycling banks.

Want to know more?

There are several in-depth guides to green and ethical shopping which you can check out or order online.
See www.thegoodshoppingguide.co.uk and www.greenguideonline.com.

FYI:
Supermarkets give away about 17.5 billion plastic bags a year; use a colourful canvas alternative.

Fashion

Nowhere is today's carbon-heavy consumer culture more apparent than in the fashion world. In 2005, UK shoppers spent £18.3 billion more on clothes than in 1993. To make new space in our wardrobes we also threw out a million tonnes of textiles. Fashion conscious? You don't have to give up new or stylish clothes. Just be more creative in your shopping habits.

Here are our tips:

Be Original – Choose Vintage and Recycled

Get in the habit of making second time round your first choice. You won't regret it. Gone are the days of cheesecloth and open-toed sandals. Instead think Patagonia's fashionable outdoor wear, branded on the basis of its recycled content; the recycled clothing snapped up by celebrity customers of LA fashion house Burning Torch; or Terry Plana's stylish "Worn Again" trainers, made with vintage shirts and salvaged leather.

Vintage clothing stores still thrive in most town centres while other entrepreneurs offer a new twist on old clothes. Junky Styling in London's East End, for example, will give your worn or badly fitting clothes a contemporary makeover (see www.junkystyling.co.uk) while online store Calico Moon offers stylish accessories from recycled materials (www.calico-moon.co.uk) and Keep & Share sells new designs knitted from old (www.keepandshare.co.uk). Or hold your own clothes swap party and invite your most stylish friends. You could end up with a new wardrobe at zero cost and zero carbon calories.

Choose **Organic**

Organic clothing made of cotton and wool are less energy intensive to produce and manufacture than clothes made of synthetic materials or doused in chemicals during production. And you no longer have to search them out on the internet or in expensive specialist stores - high street chains such as Marks & Spencer, Dorothy Perkins and Top Shop are getting in on the act. While buying local or UK made clothes should be your number one consideration, buy organic if your choice is between organic or non organic products made overseas.

Low Carbon High Street Buys

Levis offer 100% organic jeans in select UK stores. They are shipped from Hungary so the transport miles are not excessive, especially when compared with conventionally made jeans from Asia.

Check out Top Shop's organic line from Edun, the green clothes company run by Bono's wife, Ali Hewson (www.edun.ie).

Oasis offers cheap, cheerful and popular Future Organic t-shirts.

With thanks to: Green Futures magazine.

Find a **New Home** for your **Old Stuff**

Waste Is Us

Every year...

15 million people dispose of a mobile phone in the UK.
We use over six billion glass bottles and jars.

Every week...

Packaging waste dumped in UK dustbins weighs
as much as 245 jumbo jets.

(Sources: www.recycle-more.co.uk; Quaker Green Action)

Hot Tip!

For more tips and information on local services,
including recycling sites for computers,
electronics and electrical goods, check out
www.wasteonline.org.uk or call the Recycle
Now! helpline on 0845-3313131.

One person's trash is another's treasure. Charity shops, clothes and textile banks, libraries (for nearly new books), neighbourhood listserves and architectural salvage yards are all good places to offload unwanted stuff. Then of course there's eBay – otherwise known as the world's biggest car boot sale – and other online auctioneers. Make sure, though, that you sell to someone in the UK or the shipping emissions might outweigh the re-use benefits. For stuff that's really beyond re-use, recycling amenity centres are the answer. Call your local council to find the nearest one.

A typical household fills one dustbin a week, releasing 1400kg of CO_2 a year. You can cut this by 30% if you recycle or pass on all paper, glass, metal and plastic products you no longer need. For example:

C Give used clothes to charity 93kg *a year*
  Recycle your daily paper 34kg *a year*

FYI:
Every tonne of paper recycled saves 17 trees.

127

Holiday
Diet

OK, we know holidays aren't the time you want to be thinking about diets – of any kind.

But now you've bought this book, it's impossible to ignore the climate impacts of cheap flights to packaged paradise.

With opportunities for global travel never greater, carbon emissions from international flights by Britons rose by 85% between 1990 and 2002. Domestic air travel (often cheaper than the train) reached six billion passenger miles in 2002.

No surprise, then, that air travel makes up a third of the average Briton's direct carbon emissions.

To make matters worse, the climate impact of flying is proportionally greater than that of any other personal activity. This is because carbon dioxide emissions have a greater warming effect when they mix with other greenhouse gases in the upper atmosphere. Just one return trip to LA will pile 5,216kg/CO2 on to your figure, 50% of the average Briton's annual emissions.

We're not going to play Scrooge and tell you to give up your hard-earned holidays. But we will show you how to cut down those air miles and find more climate-friendly destinations. So you can travel with a clearer conscience.

FYI: on present trends, UK air passenger numbers will more than double from 188 million in 2002 to 400m a year by 2020.

Cut Your Air Miles

Let the train or coach take the strain...

Almost half of all flights over Europe cover less than 300 miles or 483 kilometres. While jumping on a plane may have become second nature, this is a distance easily covered by train unless you're literally taking a three day break. Journey times can be less of a disadvantage than you think, especially when you factor in schlepping to the airport and endless security checks. And the carbon-busting benefits are huge. Flights from London to Paris or Brussels generate ten times more carbon dioxide emissions than taking the train, according to independent research for Eurostar.

From autumn 2007, the UK's first high-speed (186mph) line will open on the Eurostar routes, whisking passengers from Central London to the heart of Paris in just two hours 15 minutes. Much less than the time it takes to drive from a London suburb to Heathrow, catch a plane to the French capital and take the metro to Montmartre.

Trains can also help you explore new and adventurous destinations not easily reached by plane. Try browsing www.seat61.com, a website created by train buff extraordinaire Mark Smith. It gives you train links across and beyond Europe, including far-flung destinations such as Ukraine, Syria, Algeria and Vladivostock.

FYI: A train passenger from London to Paris generates enough CO2 to fill a Mini; an airline passenger would fill a double-decker bus

Carbon Savings to Holiday Hotspots

Trip from London by train/bus instead of plane	C Savings:
Take Eurostar to Paris (or Brussels)	222kg
Take the train to Prague (2072km return)	756kg
Or Barcelona (2292km)	837kg
Or Edinburgh (635km)	232kg
Take a coach/ferry to Dublin (469km)	145kg

Sources: Defra, Eurostar, DfT

Our at-a-glance guide to popular destinations, above, gives you an idea of the carbon calories you can cut by changing your holiday travel habits.

To work out your personal savings to any destination, all you need to do is one quick calculation. For every 1km you travel, simply count 0.31kg/CO2 saved if you take the coach and 0.37kg CO2 saved if you take the train.

Make a "Fly Less" Pledge

As a basic rule of thumb, the fewer flights you take the better. Start with a realistic commitment, perhaps cutting long haul flights (outside Europe) to one every two years instead of one a year. Or limit yourself to two budget airline trips a year to European cities if you normally take three or four. You could always substitute a planned week abroad for two long weekends exploring Britain by train or car. If you love to shop in New York, why not cut out the jet lag and security checks, save lots of cash and carbon calories and order from Bloomingdales and Tiffany's online?

C Cut out one return short haul flight a year   = 405kg
    Cut out a return flight to the US East Coast   = 3317kg
    Cut out a return flight to the US West Coast   = 5216kg

To help keep up your resolve why not make an inspirational sticker like the one below for those challenging moments when bargain holiday ads appear on your TV or computer screen?!

If you don't know the length of your journey, ask your travel agent or visit www.indo.com/distance which calculates distances between cities worldwide, using US Census data. For mileage between UK destinations, use the AA's calculator at www.theaa.com/travelwatch/planner_main.jsp. At the end of the month, or year, tot up your total savings and add them to the Diet Masterplan under Holiday Diet. **Over to you!**

Pay Guilt Money: Offset Your Air Miles

Carbon offsetting may sound technical, but it's hitting the high street as concerned consumers seek to "cancel out" their impact on climate change. Most schemes work by paying towards energy-saving projects in developing countries, such as installing wind turbines or rooftop solar panels. Travel offsetting, especially by air, is generating the most business.

Offsetting is a feel good move rather than a real weight-reducer, like substituting saccharine for sugar rather than cutting out sweet foods. But it's better than doing nothing: we suggest that you check out these options, but don't deduct any payments from your carbon calorie total.

Book with the Travelcare high street chain (owned by the Co-operative Group) and offset your flights when you buy tickets. Eight price bands reflect the distance travelled – £3 to mainland Spain, £50 to Australia. Customers receive feel-good luggage tags which help spread the word.

Next time you fly BA, check out their carbon offset scheme at www.ba.com/offsetyouremissions. A return flight to Johannesburg will set you back £13.30.

Climate Care, one of the most established offset companies, partners with both Travelcare and BA and also takes individual donations at www.climatecare.org.uk. Peruse their overseas projects and get payment estimates for your travel plans over the next month or year.

FYI: Celebrity offsetters include Dido, Coldplay and Leonardo di Caprio.

My/Our Diet Masterplan

Take 20 minutes a month to fill this in and tot up how fast
you're shedding carbon kilos. Track your progress over
a year, deducting the monthly carbon weight loss off your
personal or household total.
1000 kilograms (kg) = 1 tonne; 1,000 grams = 1 kg.

Note: monthly carbon weight loss (kg CO2) figures have been rounded up/down
to the nearest whole figure and so may not be exactly 1/12 of the yearly saving.
If you're curious about the sources for our figures, see the explanations on page 210.

My Diet Plan	Monthly Weight Loss (Kg/CO2)	Yearly Weight Loss (Kg/CO2)	Jan	Feb	Mar	Apr	May	Jun	Jul	Aug	Sep	Oct	Nov	Dec
My Carbon Savings														
Home Diet														
Fit one energy saving light bulb	3kg	40kg												
Switch off idle electronics and appliances left on standby	13kg	153kg												
Unplug idle computers at night and w/ends	12kg	145kg												
Don't overfill the kettle. Boil what you need	4kg	48kg												
Turn down washing machine to 40C from 60C/90C	3kg /6kg	40kg /71kg												
Halve number of washes and lower your wash temperature	8kg	101kg												
Run dishwasher on an Economy setting	4kg	48kg												

My Diet Plan	Monthly Weight Loss (Kg/CO2)	Yearly Weight Loss (Kg/CO2)	My Carbon Savings											
			Jan	Feb	Mar	Apr	May	Jun	Jul	Aug	Sep	Oct	Nov	Dec
Home Diet														
Run dishwasher on an Economy setting and halve the number of times you put it on	8kg	101kg												
Run the tumble dryer half as often	13kg	156kg												
Forget the tumble dryer, rediscover the clothesline	26kg	311kg												
Replace old fridge freezer with an A rated model	15kg	180kg												
Fit a low flow shower head (family of four)	17kg	200 kg												
Turn hot water down to 60°C	12kg	145kg												
Turn heating down 1°C	25kg	300kg												
Treat your hot water tank to an insulating jacket	13kg	150kg												

My Diet Plan	Monthly Weight Loss (Kg/CO2)	Yearly Weight Loss (Kg/CO2)	My Carbon Savings											
			Jan	Feb	Mar	Apr	May	Jun	Jul	Aug	Sep	Oct	Nov	Dec
Home Diet														
Install a save-a-flush in your loo	0.048kg	0.58kg												
Spring cleaning ideas: Draught-proof windows and doors	12kg	140kg												
Seal holes in floors and skirting boards	10kg	120kg												
Fit foil behind your radiators	4kg	51kg												
Insulate your loft	125kg	1.5 tonnes												
Install double glazing	57kg	680kg												
Upgrade to a condensing boiler	83kg	1 tonne												
Garden Diet														
Swap a water butt for the garden hose	0.05kg	0.6kg												
Compost food and garden waste	23kg	280kg												

My Carbon Savings

My Diet Plan	Monthly Weight Loss (Kg/CO2)	Yearly Weight Loss (Kg/CO2)	Jan	Feb	Mar	Apr	May	Jun	Jul	Aug	Sep	Oct	Nov	Dec
Transport Diet														
Ditch the car for journeys of 2km or less - walk or cycle (five times a week)	7kg	86kg												
Take the train/bus/ Underground to work, not the car (12km commute)	(train) 51kg (bus) 57kg (tube) 58kg	607kg 688kg 690kg												
Cycle to work	71kg	854kg												
Re-think the school run: walk/bike rather than drive (4km trip, twice a day)	43kg	513kg												
Or...walk/bike 3 times a week	26kg	308kg												
Be a smooth driver: avoid sharp braking or acceleration	1.8kg (per 10km)													

My Diet Plan	Monthly Weight Loss (Kg/CO2)	Yearly Weight Loss (Kg/CO2)	My Carbon Savings											
			Jan	Feb	Mar	Apr	May	Jun	Jul	Aug	Sep	Oct	Nov	Dec
Consumer Diet														
Most rubbish comes from food and other packaging. Recycle glass, paper, card, plastics, cans.	35kg	420kg												
Recycle your daily paper	3kg	34kg												
Do one weekly supermarket shop, not three	4kg	52kg												
Buy British - a basket of UK grown food rather than produce flown from abroad	60kg													
Buy a kg (2lb punnet) of British strawberries rather than Californian	13kg per kilo bought													
Buy one kilo of British pears, instead of South African grapes	14kg per kilo bought													
Buy one kilo of British green beans, not Kenyan	10kg per kilo bought													
Buy a bottle of French wine instead of a New Zealand vintage	0.068kg per bottle bought													

My Diet Plan	Monthly Weight Loss (Kg/CO2)	Yearly Weight Loss (Kg/CO2)	My Carbon Savings											
			Jan	Feb	Mar	Apr	May	Jun	Jul	Aug	Sep	Oct	Nov	Dec
Holiday Diet														
Take the train for a UK or short overseas trip, (350km x 2), not the plane	256kg													
Or…take a coach, not the plane	217kg													
Take Eurostar, not the plane, from London to Paris or Brussels	222kg													
Take the train rather than fly London to Edinburgh	232kg													
Take 1 rather than 3 short haul (500km x 2) flights a year	810kg													
Advanced Diet														
Install cavity wall insulation	100kg	1.2 tonnes												
Switch to a green electricity supplier	117kg	1.4 tonnes												

141

My Diet Plan	Monthly Weight Loss (Kg/CO2)	Yearly Weight Loss (Kg/CO2)	**My Carbon Savings**											
			Jan	Feb	Mar	Apr	May	Jun	Jul	Aug	Sep	Oct	Nov	Dec
Advanced Diet														
Install 1-1.5 kilowatt rooftop wind turbine for DIY electricity	47kg for 1Kw 95kg for 1.5Kw	568kg for 1Kw 1.1 tonnes for 1.5Kw												
Install rooftop solar PV panels for DIY electricity	76kg	909kg												
Buy a hybrid car	0.064kg /km	1 tonne every 16,000km												
Sell your car, use public transport	142kg	1.7 tonnes												
Give up beef	5.2kg CO2 per 0.45kg (1lb)													

My Monthly Weight Loss Totals

Jan	Feb	Mar	Apr	May	Jun	Jul	Aug	Sep	Oct	Nov	Dec

Total Weight Loss for Year ....

Total Weight Loss...the fun part!

You can calculate this for a year or part year, whenever you feel ready…

1. My Old Weight: .... kg CO_2

2. My Total Savings:
Diet Masterplan   = ... kg CO_2
Additional actions   = ... kg CO_2

3. My New Weight: .... kg CO_2
(deduct 2 from 1)

You see things and say: " Why? "; but I dream things that never were and say " Why Not? "

George Bernard Shaw
Playwright and philosopher

So here you are, looking slimmer, feeling fitter and having shed kilos off your carbon weight by following some of the simple advice in the previous sections. You've got the bit between your teeth and you're ready for a bigger challenge.

This section - our advanced low carbon diet - is for people who are prepared to invest more time and (sometimes) more money on lifestyle changes. Also try it if you just want to dip your toe in greener waters and see how it feels...

There's a bit more detail to digest and a bigger commitment required of you. But remember, super dieters, a bigger commitment translates into bigger savings for your carbon calorie count and your wallet. Not to mention a healthier, more sustainable planet.

Advanced **Home** Diet

Let's start with what you can do at home to reach carbon nirvana.

Energy saving makeovers are increasingly common in both draughty old homes and new buildings. Depending on your inclination and budget, you could either opt for a step by step approach, perhaps installing heavy duty insulation one year and off the shelf solar panels (now sold at B&Q) the next. Or you could go the extreme makeover route, adopting most or all of our suggestions in one go. This would cost quite a bit upfront, but you'll be paid back handsomely in dramatically reduced bills over a few years.

Whichever route you choose, seek free, comprehensive and expert advice from your local Energy Advice Centre on what would work best for your particular circumstances (call 0800 512 012 for details).

Here goes!

Step 1: Jumping ship: switching to Green Energy

Switching to a green energy supplier is a sensible (and pain free) first step which at a stroke cuts a tenth off the average householder's annual carbon weight. All you do is opt to have your electricity generated from a renewable source – wind, hydro-electric, solar, biomass (wood and fuel crops) or wave power – rather than from a fossil fuel such as coal or oil. You may not even have to change supplier.

Most major power companies now offer customers 'green tariffs' or 'green funds'. A green tariff will match your electricity use with energy from renewable sources – so for every unit of electricity you buy, the company will deliver a unit of renewable energy to the national grid. A green fund uses money from your bill to support renewable energy or other environmental projects or research. To make sure your money is spent directly on supplying renewable energy, choose the green tariff option.

FYI:

3% of UK electricity comes from renewables. The government aims to increase this to 10% by 2010, 15% by 2015.

Switching to a green tariff won't make any difference to the way electricity feeds into your home or how you are billed. Nor will it cost an arm and a leg. Bills are the same or just a few pounds more than conventional supply. Go to www.greenelectricity.org/ or www.uSwitch.com to compare different suppliers in your area.

C 1.4 tonnes *a year*

One small reminder: don't make this switch an excuse for not continuing to cut back on your electricity use as much as possible. **Doing both is best.**

Hot Tip!

Before signing on the dotted line, make sure the supplier is providing you with 'green' electricity over and above its legal obligation to supply almost 5% of its output from renewable sources.

Step 2: Mind the gap: cavity wall insulation

If your home is uninsulated a third of the heat you pay for escapes through the walls, along with the carbon dioxide pumped out by your heater. If you have cavity walls, as do most houses built after the 1920's, this is easy to fix. An installer will simply inject mineral wool or polystyrene beads between the two layers which make up your outside walls.

The whole process takes three or four hours for a three bedroom semi-detached house. And there's no mess; the insulation is injected from the outside. If you get a grant, the average installation costs around £260 and pays for itself in lower heating bills in under two years. Go to www.est.org.uk/myhome/gid/ to investigate grants in your area and to your local Energy Advice Centre (Tel: 0800 512 012) to find an installer.

OK, so maybe it doesn't make for good dinner party conversation, but apart from that what's not to like?

C 1.2 tonnes *a year*
£ 130 - £160 *a year*

Step 3: Carbon super-savers: gadgets and gizmos

Eco-kettle

This jazzy new kettle releases only the water you need into a boiling chamber when you choose from 1 to 8 cups. In consumer trials, it used 30% less energy than standard kettles (www.ecokettle.com).

Low-flow showerheads

These mix water with air to maintain pressure and cut water use by half. If it takes less than 20 seconds to fill a one-gallon bucket using your shower, you need one of these.
Or try using a shower timer to cut back your ablutions by a minute or two each day (www.rippleproducts.com/shop/productsearch).

Standby smart plug

Instead of constantly bending over to push off standby buttons, invest in an intelligent plug that does it automatically when computers and TVs are switched off. Order from the Centre for Alternative Technology at www.cat.org.uk/shopping/ under Practical Solutions.

Powermeter

This handy plug-in meter shows the electricity consumed by an individual appliance and the cost of the electricity used (web address as above).

Wind-up torch

This uses wind-up technology to power a 10mm ultra bright light emitting diode or LED – the kind of bulb you now see cyclists using. A mere 60 seconds of elbow grease provides 1 hour of shine time. See www.cat.org.uk/shopping/ under Household Products.

Step 4: Move towards off-the grid living: install a solar panel or wind turbine

Setting up your own mini renewable power station by installing solar panels, a wind turbine or a wood pellet stove is the ultimate way to cut your contribution to climate change. You'll also increase the value of your home, enjoy a healthy dose of good karma and be up there with the celebrities on next year's must-have planet-friendly accessory.

These technologies aren't cheap and may take some years to pay back their cost. Still, demand is now high enough that B&Q, the DIY superstore, is selling budget-priced wind turbines and solar panels in 300 stores at £1498 each.

The government is also doing its bit, providing homeowner conversion grants through the Department of Trade and Industry's Low Carbon Buildings Programme. See www.lowcarbonbuildings.org.uk or call 0800 9150990. North of the border, grants are available from the Scottish Community and Householders Renewables Initiative at www.est.org.uk/schri (Tel: 0800 138 8858).

Choice 1: Sun Harvesting

Two very different technologies are used to tap into solar power: solar water heating and solar electricity.

A solar water heating system works by pumping water through a flat plate positioned on a south-facing roof and then feeding it into a hot water cylinder. Combined with a conventional water heater, it can provide about half your year-round hot water needs. Apart from the feel-good factor, a solar water heating system doesn't need much maintenance and should last for decades.

A professionally installed system for a typical house will cost £2,000-£3,000. Cut your costs with a grant of up to £400 from the DTI's low carbon buildings programme.

C About 307kg *a year, depending on the fuel replaced*
£ 40 *a year*

Hot Tip!
Check whether you need planning permission to install rooftop solar panels.

FYI:

there are 78,470 solar water systems up and running in the UK and 1301 solar PV installations.

Solar electricity uses 'photovoltaic' (PV) solar panels to convert energy from daylight into electricity to run appliances and lighting. The cost will give you pause – about £12,000, but a DTI grant can cut this by half. A 2kW PV panel could generate half the electricity you need to light your home and power your appliances. Systems connected to the national grid need little maintenance – just keep the panels relatively clean and watch that heavy shade from trees is not blocking out the light. The wiring and components should also be checked regularly by a qualified technician.

C 909kg *a year*
£ 160 *a year*

Want to know more?

See www.est.org.uk/myhome/generating/ and www.solarcentury.com or call the Centre for Alternative Technology free on 01654 705 989 (www.cat.org,uk). For government approved installers visit www.lowcarbonbuild-ings.org.uk under Information and Resources.

Choice 2: Wiring up the wind

Wind power is very high profile, especially since David Cameron purchased a turbine. It also makes sense. The UK has 40% of Europe's wind power, yet we use only 0.5% of it to generate electricity.

But in practice wiring up your home to the wind is generally less easy than going the solar route, depending on your property's location. The majority of wind turbines are attached to a tall mast and are most suited to off-the-grid (i.e. very remote) locations, although building-mounted systems are starting to come on the market. You also need to live in a reasonably windy area and a location where trees and power lines do not present obstacles (not always easy in cities).

To see if your home passes these tests, get advice from a government approved installer listed at www.lowcarbonbuildings.org.uk or telephone the DTI information line on 0800 915 0990. If you get the green light, a 1kW turbine will supply one third of your electricity needs while a 1.5kW model will deliver 71% of supply. Small home turbines cost between £1,500 and £5000, but DTI grants contribute up to 30%.

☙ 500-1000kg *a year, depending on size of turbine*
£ 100-£200 *a year.*

Want to know more?

See www.est.org.uk/myhome/generating/types/
wind/, the British Wind Energy Association
at www.bwea.com, or the Centre for Alternative
Technology as listed under solar advice.

FYI:
Number of micro wind
turbines in the UK: 650

Hot Tip! You will need planning
permission to install a wind turbine.

Choice 3: Carbon supersaver stoves and boilers

Known as biomass stoves, these stand-alone 6-12 kilowatt appliances burn wood pellets (made from wood industry waste), wood chips or logs to provide space heating.

Burning biomass is "carbon neutral", which means it adds no net carbon emissions to the atmosphere. Although burning wood products does release carbon dioxide, the same amount of CO_2 was absorbed while the tree was growing. To keep the cycle carbon neutral, make sure your pellets or wood chips come from a woodland where replanting matches harvesting.

Most biomass stoves heat a single room but larger models can be fitted with a back boiler to supply hot water and even central heating through radiators. They come with automatic feeders, so just sit back and enjoy. Stand-alone stoves – about the size of an ordinary wood-burning stove – cost £1500-£3000 and wood-fuelled boiler systems around £5000 for a 20kW installation, enough to heat a three bedroom house. The DTI offers grants of up to £600 for a stove and £1500 for a boiler system. Fuel costs extra.

C 2.6 tonnes a year for a wood-fuelled boiler system providing all space heating and hot water needs.

FYI:
There are 150 biomass boilers in the UK. Go on, be a pioneer!

Seek Out a Smart Meter

Currently being trialled by several UK electricity companies, this new breed of 'smart' meter is poised to offer a valuable slimming tool for the keen carbon dieter. Smart meters are compact, interactive devices, with a handy digital display that tell you how many kilowatts of power you are using at any one time and work out the daily and monthly costs to your household. If you turn off a piece of electrical equipment, the meter will instantly adjust the readings. So you'll easily be able to work out, for example, how much it costs to keep a TV on standby or your hall light on all day. To check out a state of the art smart meter go to www.moreassociates.com/energy/. To find out if there's a trial in your area, contact your electricity supplier.

Want to know more?

See www.est.org.uk/myhome/generating/types/ or the CAT website and advice line.
For installers and installations approved for a government grant, go to www.lowcarbonbuildings.org.uk.

Advanced **Garden** Diet

The advanced carbon dieter surveys a garden and sees not a lawn to lounge on or flowers to admire. He or she sees carbon saving potential under every bush. By growing your own herbs, vegetables and fruit, planting trees and wild flowers and using recycled materials to spruce up your outdoor décor, you can enjoy some rapid weight loss. So let's get started.

DIY food: zero carbon fruit and veg

In our Consumer Diet, we urged you to buy in season food grown locally or within the UK. Now how about taking the next logical step and growing some of your own food? It's not as hard or as time-consuming as you might think, will keep you fit and produce a ready supply of fresh food. Plus you'll be cutting out food packaging and food miles with one easy step – literally – into your back garden.

Start by growing something straight forward and popular in your home – perhaps onions, potatoes or lettuces. Put a few willow cones in a flowerbed and grow beans or cucumbers over them. Invest in a book or two and ask gardening friends to share their experience. Digging out weed roots in your new vegetable beds will be a chore. But once that's done it should be plain sailing, especially if you use rich home-grown compost to nourish the soil and mulch to suppress weeds. **To help your planning, here's a seasonal planting and harvesting calendar:**

The Carbon Dieter's Seasonal Food Calendar

Spring

Plant out: lettuces, beans, peas, onions, leeks, spinach, carrots, potatoes and beetroot. In late spring, add french and runner beans, courgettes and marrows. Cover young plants with sawn-off plastic bottles to protect them from slugs and early spring frost.

Summer

Plant out: tomatoes, cucumbers, winter cabbages and cauliflower.

Enjoy: lettuces, carrots, tomatoes, cucumbers, beans, courgettes, spinach, celery, rocket, peas, onions, raspberries, strawberries, blackcurrants.

Autumn

Plant out: rhubarb and prune your fruit trees.

Enjoy: apples, pears and potatoes (or dig up and store in paper sacks for year round use).

Winter

Plant out: garlic and shallots. Buy potato seeds, store in old egg-boxes in a cool, frost-free spot (the garden shed or a cellar for example) and leave to sprout.

Enjoy: early brussel sprouts, potatoes, parsnips, cabbage.

C 48kg *if you grow 4kg of beans rather than buy them from abroad*

C 45kg *if you grow 3kg of strawberries rather than buy them from abroad*

Plus, save 50 litres of water every time you use lettuce and herbs from your garden instead of a bag of ready washed supermarket salad.

Want to know more?

Try the Henry Doubleday Research Association at www.gardenorganic.org.uk/; the Royal Horticultural Society at www.rhs.org.uk/advice/calendar.asp; and www.bbc.co.uk/gardening/htbg/module7/.

Go wild about flowers

Buying cut flowers, especially exotic overseas varieties, is a real no-no, because of transport-related emissions. Either buy local or simply plant some bulbs in your garden and fill your vases with cuttings. You could also try planting a wildflower patch – think poppies and ox-eye daisies – which will only need watering once a month.

Hot Tip!

Buy packets of wildflower seed mixed with grasses guaranteed to be local native species harvested from British meadows.

Plant trees

It's a simple equation. Trees absorb carbon dioxide as they grow. So, by planting more trees, we can reduce our carbon footprint, as long as we plant new trees when the old ones die. Planting a fruit tree will deliver both food and carbon offsetting – two for the price of one. Another good choice is to plant tree saplings that withstand dry conditions like pine, eucalyptus, cypress, holly and cedar.

C 1 tonne *absorbed by an average tree by the time it reaches maturity.*

FYI:

the world's trees absorb around seven million tonnes of carbon a year, equivalent to storing 26 million tonnes of carbon dioxide.

Green your garden decor

Go solar: sun-powered garden lights don't just cut carbon calories; they remove the need for wiring. The batteries simply charge up in daylight and start to twinkle at dusk, although some come with a manual over-ride switch. Be creative: mark pathways and garden features with a gentle glowing light.

Use recycled materials: for seating and fencing, try recycled plastic or reclaimed wood. For patios, think about reclaimed stone and for statuary, visit an architectural salvage yard. For some inspiration, go to www.wilsonsyard.com or see what you can do with recycled plastic at www.buildingdesign.co.uk/arch-1/marmax-products/plastic-fencing.htm

No garden, no problem

Growing your own food doesn't require acres of land; a balcony or even a broad windowsill will do. Plant cooking or salad herbs in pots (basil, thyme, rosemary and sage all do well). Or show off an unusual window box sprouting carrots and spring onions. If you have a patio, you're halfway there. Grow vegetables in containers, using dwarf varieties suited to small spaces. Don't know your dwarf variety from Jack's beanstalk? Here are some suggestions to get you started:

Cucumber: Salad Bush Hybrid, Spacemaster, Bush Pickle

Aubergine: Bambino, Slim Jim

Green Beans: Blue Lake, French Dwarf

Green Onions: Crysal Wax, Evergreen Bunching

Leaf Lettuce: Buttercrunch, Salad Bowl, Bibb

Peppers: Frigitello, Cubanelle, Robustini

Tomatoes: Patio, Pixie, Tiny Tim, Spring Giant, Small Fry

If you literally have no outdoor space, investigate nearby allotments. These have fallen from favour since the days when 1.5 million of them supplied much of the fresh produce eaten by less well off families. Of the 250,000 or so remaining, up to a fifth are lying vacant. To find out if any are near you, ring your local council.

Want to know more?

Read the Allotment Keeper's Handbook by Jane Perrone available from Guardian Books.

Advanced **Transport** Diet

Let's assume you've followed our advice in the Transport Diet and are driving smarter and less. Well done. But you still own the same car and it drinks up fuel. What can you do to turn those kilos you are shedding into tonnes? We have three options for you to consider: buy a hybrid; explore funky fuel choices that emit less carbon dioxide; or (deep breath, needed here) sell those costly wheels and join a car share club.

Do these and you won't just be super carbon dieting; you'll be helping launch a trend likely to pay handsome dividends in the near future. David Miliband, the Environment Secretary has talked openly of the need for carbon rationing cards for personal fuel use. London's mayor, Ken Livingstone, recently hiked the congestion charge to £8 a day and from 2009 this will increase to £25 for gas guzzlers – large estate cars and 4 x 4s. Our suggestions may seem expensive or inconvenient now. But imagine a few years ahead when alternative fuels are dirt cheap or your neighbours are queuing up to buy your unused or under-used carbon credits. You'll be the one smiling then.

Buy a Hip Hybrid

Proud (mostly Prius) owners include:

> David Milliband
> Prince Charles
> Gwyneth Paltrow
> Euan McGregor
> Cameron Diaz
> Leonardo DiCaprio

UK public: about 5,000 hybrids bought in 2005

US public: 26,000 hybrids bought in August 2006

How Do They Work?

While hybrids are increasingly popular in the UK, Europe and the US (Toyota alone has sold 750,000 Prius models worldwide) misconceptions still abound.

Hybrids run on "dual fuel", using power from an electric battery at low speeds, in cities and towns, and switching automatically to a petrol engine on faster roads. Forget those mental images of plugging your car battery into a socket in your garage. The electric battery recharges whenever you brake, with no other action needed on your part. To keep the car running all you have to do is fill up with unleaded at any petrol station in the country. If you drive a standard petrol car and do an average annual mileage you should cut fuel use by a third over a year.

Which to Choose?

Toyota Prius:

The Toyota Prius won Car of the Year at the 2005 Detroit and Paris motor shows and in 2006 What Car? magazine hailed the latest model as Britain's "greenest family car". It coasts along city streets in near silence and switches to a 1.5-litre petrol engine on faster roads.

Pros: it emits just 104g/km of CO_2, the lowest of any car on sale; great mileage at 67.5mpg; £30 road tax and no congestion charge for Londoners.

Cons: it's pricey compared with similar petrol cars – £17,545 and up. It will take you several years to make up the difference in lower fuel bills.

Honda Civic Hybrid:

The Prius's main rival, the Civic saloon, is smaller and (a bit) cheaper with a 1.3-litre petrol engine. The latest model was named Britain's "greenest small family car" by What Car?

Pros: great for fuel economy and low emissions – 61.4mpg and 109g/km CO2; £30 road tax; "unusual parts" are guaranteed for the car's life; no London congestion charge.

Cons: price – from £16,300 compared with £13,000 and up for a standard Civic.

Lexus GS450 (luxury saloon) and RX400 (4x4):

For those who want to keep driving luxury cars or four wheel drives but feel guilty about it (David Miliband owns the saloon). You get more mpg than rival petrol-only models and all the mod cons, but at a price; namely £36,000 and up.

Pros: more carbon-lite than driving a petrol only equivalent.

Cons: £180 annual tax disc; still high carbon calories – 186g/km for saloon, 192gm/km for 4x4.

C 0.064kg/km or 1 tonne per 16,000km if you swap an average-sized petrol car for a hybrid Honda Civic.

C 2.2 tonnes a year if you trade an SUV for a hybrid

Want to know more?

www.whatcar.co.uk, see Help and Advice; www.green-car-guide.com; www.eta.co.uk

P.S. Reward Carbon-Conscious Companies…

If, after doing your sums, you're not yet ready or able to invest in a hybrid, don't beat yourself up. The costs should come down as supply increases. If you buy a new (or second hand) car in the meantime, revisit the tips on carbon lite models in our standard Transport Diet. All European manufacturers have agreed to cut the average CO_2 emissions for their new cars to 140 grams per kilometer by 2008. So, in theory, all new models should be cleaner and greener than their predecessors. However, some manufacturers – BMW, Volvo, Audi, Nissan, Mazda, Suzuki – are really dragging their feet; while others – Fiat, Citroen, Renault – have already met the target. So why not use your buying power to reward the good guys and spurn the bad?

Try Clubbing It

This involves two steps into the unknown a. selling your car and b. joining a car club. Sounds scary, we know. But before you rule it out, think about what you could save. And we don't just mean carbon calories. We mean hundreds, perhaps thousands of pounds a year.

The less you drive, the higher the costs you pay, per kilometre, to own your own car. The AA Motoring Trust produces a handy table which calculates that medium sized petrol cars cost their owners £2,600 a year in depreciation, insurance, road tax and breakdown cover, if driven a typical 16,000km (10,000 miles). Once you add in running costs, including petrol, the figure rises to £4087. (The calculator assumes you buy a new model every four years and also gives figures for diesel models.)

That's a lot of money. So why not indulge us and spend a minute on the following steps:

First work out exactly what your car is costing you by using the AA running costs table at http://www.theaa.com/allaboutcars/advice/advice_rcosts_petrol_table.jsp.

Next make a list of journeys you make every day or week and see how you could cut them down. Can you walk or cycle some? Are there bus or underground routes you could use? Could you use delivery services instead of driving to the supermarket or takeaway restaurant? How about shopping online instead of at the shopping centre? Where car trips are unavoidable, could you make them all in one day a week, fortnight or month?

Now work out how much you would save by walking, cycling or using public transport whenever possible and renting a car club vehicle to cover your essential driving needs at £5 an hour plus petrol costs. We bet you'd be quids in by ditching the car, unless you drive well over 10,000 miles a year.

The Lowdown on Car Clubs

Car clubs rent vehicles to members for £3-5 an hour (24 hour rates can be cheaper) plus petrol costs (the first 30 miles are usually free). They already operate in 26 UK towns and cities, boast 21,000 members and are expanding rapidly. Membership is generally free or costs no more than £50. Clubs can usually provide a car parked within ten minutes walk of your home or office.

Once you've joined, simply book a car over the phone or online, use your membership smart card to unlock the doors, enter a pin number in the dashboard and retrieve the keys from the glove compartment. Insurance is covered and members are usually billed once a month. The three main operators are Streetcar in London, Brighton and Southampton; CityCarClub, which has 11,000 members in 11 cities and WhizzGo, whose London cars are exempt from the congestion charge as they run on liquid gas (see right).

What Could I Save?

Say you drive a standard petrol car 10,000 miles a year (the national average). According to the AA website, your total annual motoring costs, including petrol, would be £4087. If instead you joined a car club and rented a vehicle twice a week, for six hours total, 46 weeks a year, your annual motoring costs (at £5 an hour) would be £1840 - a difference of £2247.

Even if you took a couple of UK train trips and a cab ride or two a month your savings should still top £2000. And you'd make money selling your car. So think about it. Just imagine what you could do with that money…

Want to know more?

See www.streetcar.co.uk,
www.citycarclub.co.uk and
www.whizzgo.co.uk.
Smaller, not-for-profit clubs are listed
at www.carplus.org.uk

Explore Funky Fuels

Alternative fuels which burn less carbon or none at all have been around for a while, but have so far failed to enter the mass market. Here's a quick guide:

Funky Fuel 1: Liquefied Petroleum Gas (LPG)

A carbon-lite mixture of butane and propane available at 1400 (1 in 9) UK filling stations. A few Vauxhall, Volvo, Ford and Nissan models can be bought LPG-ready. Or you can convert your car for around £1500.

Pros: costs half the price of petrol and produces 10% less CO_2 and zero soot emissions.

Cons: gets fewer mpg than petrol or diesel cars; suppliers are scarce.

To find your nearest LPG filling station and search for a contractor to convert your car, visit the Energy Savings Trust website at http://www.est.org.uk/fleet/Vehicles/Alternativefuels/Alternativefuelsrefuellingmap/

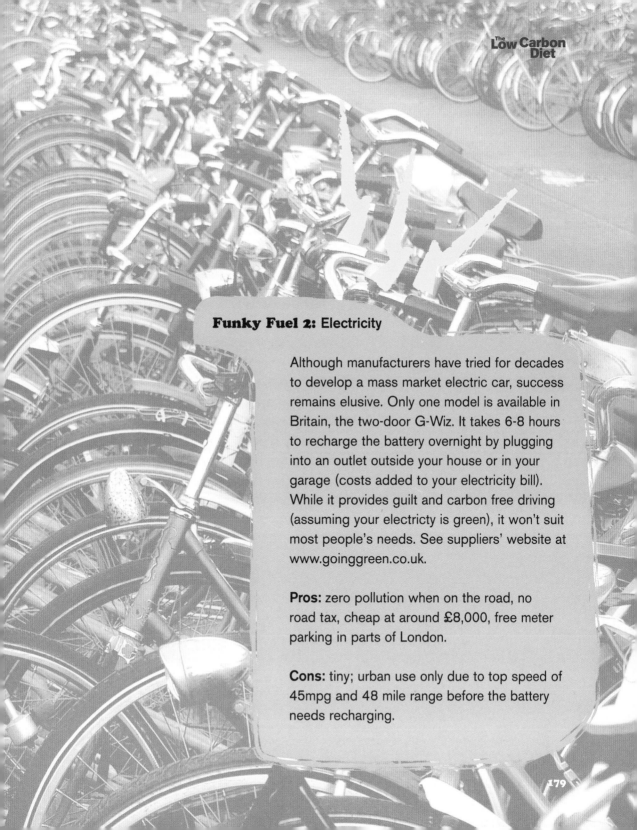

Funky Fuel 2: Electricity

Although manufacturers have tried for decades to develop a mass market electric car, success remains elusive. Only one model is available in Britain, the two-door G-Wiz. It takes 6-8 hours to recharge the battery overnight by plugging into an outlet outside your house or in your garage (costs added to your electricity bill). While it provides guilt and carbon free driving (assuming your electricty is green), it won't suit most people's needs. See suppliers' website at www.goinggreen.co.uk.

Pros: zero pollution when on the road, no road tax, cheap at around £8,000, free meter parking in parts of London.

Cons: tiny; urban use only due to top speed of 45mpg and 48 mile range before the battery needs recharging.

Funky Fuel 3: Biofuels

These run on waste vegetable oil, wheat, rapeseed and sugar cane. As long as the crops are replaced they are zero emission, because even though they emit CO_2 when burned in a car, the crops absorbed the equivalent amount while growing. Standard diesel cars can be converted to run on biodiesel. A few Ford and Saab models, including the Ford Focus Flexi Fuel Vehicle and the Saab Biopower will run on E85, a blend of 85% bioethanol and 15% petrol, as well as regular petrol.

Unfortunately, UK service stations, most of which are run by the oil industry, are proving reluctant to open up their forecourts. For now, you can only fill up at 150 locations.

Pros: fewer emissions than petrol or diesel when a mix of biofuel and petrol is used – under 100g/km CO_2 for the Ford FFV; slightly cheaper per litre than petrol.

Cons: chances are there's no filling station within 50 miles of you.

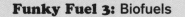

FYI: in Sweden, thanks to duty free incentives, Ford's flex-fuel models are outselling its new petrol and diesel cars.

Advanced **Consumer** Diet

Give up Beef or Try a Meat-Free Diet

Remember how eating less meat was one of the five golden rules for our food diet? If you've done this without too much trouble, congratulations. Now, how about taking it a step further by giving up beef altogether or trying out a vegetarian diet? According to the Climate Outreach and Information Network (COIN) each 1.1kg (1lb) of meat produced, transported and refrigerated, produces on average greenhouse gases equivalent to 4.7kg of carbon dioxide (5.2kg in the case of beef). *Think of it as a chance to try new tastes and broaden your cooking skills.*

Hot Tip!

Try to avoid stocking up on frozen produce as a lot of energy is used in refrigeration and storage.

Only Buy UK Produce

Greenhouse gas emissions produced by transporting food from plough to our plates are rising every year. At this rate, there is little chance that the government's pledge to cut the environmental costs of food transport to 20% below 1990 levels by 2012 will be fulfilled.

Doing your bit to cut down on food miles is simple: wherever possible, buy British only.

Our Advanced Garden Diet (page 164) has already provided you with a handy, comprehensive calendar of seasonal home-grown fruit and vegetables. And with most big supermarkets selling more local produce, you should be able to find what you need during your normal weekly shop.

If you have several supermarkets nearby, a recent report by the National Consumer Council, Greening Supermarkets, provides useful guidance on where to find the most choice and which shops are doing the most to promote seasonal UK food. (For more, see http://www.ncc.org.uk/responsibleconsumption/greening-supermarkets.pdf).

If you get bored with the supermarket offerings, try seeking out new, local varieties of produce, cheese, bread, honey and so on, in farmers markets and farm shops. See www.farmersmarket.net for your nearest outlet.

FYI: spring onions home delivered in a farm box produce 300 times less CO_2 than those flown 9,000 miles from Mexico and bought in a supermarket.

Be a Climate-Friendly Consumer

Surprising but true - British consumers now spend more on green and ethical goods and services (£29.3bn in 2005) than on cigarettes and alcohol. The biggest slice, according to a report from the Co-operative Bank and Future Foundation, is spent on ethical investment funds. But more of us are also buying products and choosing services on the basis of a company's record on environmental issues, especially climate change.

Carbon dieters can encourage this trend by comparing companies' records and commitments on improving energy efficiency and reducing corporate carbon emissions before signing up for services such as banking, mortgages, phone and internet access and car and house insurance. You can do this by visiting their websites and searching out environmental or corporate social responsibility reports. Or refer to the green shopping guides highlighted on page 123. To give you a headstart, check out the following:

Carbon Conscious Banks:

Online banks are good because they cut out car journeys and energy used in generating and posting statements. Among traditional banks, the Co-Operative Bank really puts its money where its mouth is. In 2005, it paid for 50,000 tonnes worth of carbon offsets on customers' car insurance, flight travel, and mortgages (see below). It also runs 99% of its 3,000 premises on green electricity contracts.

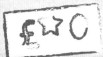

Carbon Conscious Mortgage Lenders:

The Norwich and Peterborough Building Society offers a "carbon neutral" mortgage, paying to plant trees to cancel out CO_2 emissions from your home for five years. See www.npbs.co.uk/about-us/environmental-policy.asp. The Co-Op Bank (through ClimateCare) will offset 20% of your home's energy use for the life of your mortgage. Call 0800 0288 288 for details.

Carbon Conscious Computer Products Suppliers:

Hewlett Packard produces more than 1,000 computers, printers, fax machines and other products that qualify for the US Environmental Protection Agency's Energy Star rating and logo. The company's latest 'Blade' PC is ten times more energy efficient than a typical desktop computer.

Carbon Conscious TV Providers:

Satellite TV company BSkyB is investing in renewable energy technologies and has moved its corporate taxi account to www.greentomatocars.com which uses hybrid vehicles.

Embrace Carbon Free Gift Giving

Between birthdays, Christmas and Mother's, Father's and Valentine's Day celebrations, we buy a lot of gifts for our nearest and dearest. With the energy used in manufacturing, packaging, transporting and eventually disposing of these presents, the CO_2 emissions really add up. So how about a new approach: carbon-free gift giving?

The boom in carbon offsetting for travel has produced a handy sideline in offset gifts. For example, through Good Gifts (at www.goodgifts.org) you can pay the Tree Council to regenerate a wildflower meadow or bluebell wood for someone's birthday. Through Climate Care, you can pay to offset other people's yearly CO_2 footprint (£90), their home footprint (£40) or their annual driving emissions (£10).

If this seems too impersonal, try giving the gift of time. Present a card pledging to babysit, dog walk or garden for a busy or homebound friend or relative once a fortnight. Or buy carbon lite gifts, such as solar powered toys (available from www.bynature.co.uk) or wind up radios, recycled coasters and water-powered clocks (from www.carbonneutral.com/shop/).

The **Low Carbon Diet**

Start a Chain Reaction

A clever way of offsetting your carbon calories is to buy carbon credits available under the European CO_2 emissions trading scheme. Companies generally buy these, so that they can continue to emit CO_2 and avoid updating energy-intensive production processes.

If enough individual consumers buy the carbon credits instead (through an offsetting company) they become scarcer, their price rises and companies are encouraged to take practical action to reduce their greenhouse gas emissions. For more information, visit www.puretrust.org.uk.

Advanced **Holiday** Diet

Fly as a Last Resort

This is a tough one. But you're up for our advanced diet, so here goes.

Flying really is the single most climate-destructive activity you can engage in. And unlike heating or lighting your home, it's not one of life's essentials. So how about a commitment to board a plane only as a last resort – say for family weddings and births, landmark anniversaries or other special occasions. If this seems too hard or too vague, make a pledge to limit personal flights to one every 12 or 18 months. (Business related travel doesn't count).

Of course, this doesn't mean you have to give up foreign travel. As we've already pointed out, you can reach most of Europe by train, coach or of course car. Long haul destinations are trickier, although almost anywhere is reachable by train and ferry if you have the time. (Don't believe us? Check out www.seat61.com). Or try the old-fashioned choice and take a ship.

FYI:

Global air travel generates nearly as much CO_2 a year as all human activity in Africa.

The Upside of Earthbound

In dieting terms, foregoing flights is like giving up sugar or alcohol. Hard at first to break the habit; but full of benefits that grow on you as time passes:

Avoid Airport Rage – queues, security checks, missing luggage. Who needs it?

Jettison Jet Lag – no more sleepless nights, aviation flu, swollen feet, deep vein thrombosis…

Get Two (or more) for the Price of One – stop off en route; visit Paris then Nice or Berlin on the way to Prague. You get the idea.

Save Money – perhaps not if you take the train instead of a bucket flight (although book ahead on Eurostar and you can get to Paris for 50 quid) but certainly if you stop flying long distance.

By the way, you won't be as lonely as you might think. In a 2006 YouGov poll one in ten people questioned had stopped flying or were using planes less due to climate concerns.

C 1.8 tonnes *for every long haul flight (6000km return) you give up a year.*
405kg *for every short haul flight (1000km return) you give up a year.*
4.2 tonnes *for every trip to the Caribbean (14,000km return) you don't take.*
5.7 tonnes *for every trip to Thailand (19,000km) you don't take.*

Go Active

Active holidays are all the rage, with good reason. Instead of crawling home with sunburn and a week-long hangover, you return fitter and healthier. You'll also help your carbon diet by choosing leisure pursuits that generate few, if any, carbon calories. Think walking, trekking, cycling, mountain biking and horse-riding.

Cycling: forget the dorky image, cycling is in. Colin Farrell, Brad Pitt, George Clooney and Lindsay Lohan have all recently been spotted on two wheels. Nor do you have to do a Lance Armstrong. There are plenty of gentle, picturesque routes in England, France and Italy, all of which have well-established cycling networks. The charity Sustrans, which coordinates the 10,000 mile National Cycle Network provides maps of scenic routes at www.sustrans.org.uk or call 0845 113 00 65. If you'd rather someone else did the organising, Responsible Travel offers cycling holidays accessible by train in England, Wales and Spain.

Hiking: also increasingly popular, both at home and in sunnier climes. The Sports Council publishes a national register of guided walking holidays. Check out Responsible Travel (www.responsibletravel.com) which offers walking holidays in the Scottish Highlands that set off from train stations and holidays in the Loire Valley which include Eurostar tickets. Or try horse riding in Spain or alpine mountain biking with Exodus (www.exodus.co.uk) voted most environmentally responsible tour operator at the 2006 British Travel Awards.

C 300,000 tonnes saved by all the leisure and commuter trips made on the National Cycling Network instead of by car in 2005, according to Sustrans.

Want to know more?

Try buying The Ethical Travel Guide: Your Passport to Alternative Holidays (Tourism Concern/Earthscan 2006) or Alistair Sawday's Green Places To Stay.

Park the Passport, Discover Britain

Choose Made in Britain holidays over the foreign variety and you can save serious carbon calories. Drive hundreds of miles; the impact still won't compare on the diet scales with flying to Europe. If you take the train, you'll do even better.

So take advantage of what our own small island has to offer. Madonna and Gwyneth Paltrow like it so much they live here. Why not get out and discover it?

If you plan to leave the car (and road rage) at home, check out Scenic Britain by Train and Scenic Britain by Bus for ideas, available from www.transport2000.org.uk.

Some food for thought:

Carbon Dieter's City Breaks
Oxford, Cambridge, Stratford, Bath, York, Edinburgh. All are very easy to access by train or coach and get around on public transport or on foot. They also discourage the motorist, with city centre traffic restrictions and high parking fees.

Countryside Breaks
Local authorities covering the New Forest, Hadrian's Wall, the Lake District, North York Moors and Yorkshire Dales are all part of a "car free leisure network", offering good public transport links from local towns. While you're at it, join the camping crowd. It's so in that top designer Cath Kidston has designed a flowery tent for Milletts.

Seaside Breaks
Cornwall's tourist towns run a summer "hoppa" bus service between picturesque seaside destinations while the National Cycle Network maintains several long, scenic coastal routes.

Group Dieting

It's often easier to lose weight – whether body fat or carbon kilos – when you're not going it alone. Witness the success of Weight Watchers. So, whilst most of this book is devoted to personal advice for you or your household, here we offer some suggestions for a group carbon diet.

The benefits are obvious. You will have more incentive to stick to your diet goals because of peer pressure. You can encourage others to do the same. And together your collective actions will make that much more difference.

Start a Carbon Diet Club

These could be based in your street, neighbourhood, village, church, workplace, school, book club – wherever you have the best contacts.

At Work: If you commute by car and lots of your colleagues do the same, an office based carbon club focused mainly on transport emissions could find a good audience. The simplest way to get people enthused? Organise a commuter car share roster matching colleagues who live nearby. Point out that they would save on petrol bills, cutting costs as well as carbon calories. If your workplace is small, you might try to broaden your car pool through www.liftshare.com or www.sharea-journey.com which match people living nearby. If you work in a large organisation, you may want to get personnel staff involved and spread the calorie cutting beyond your own department. The government's Energy Efficiency Best Practice Programme offers free, site specific expert advice to help organisations develop and implement travel plans. (Call 0800 585794 or visit www.energy-efficiency.gov.uk/transport.).

At School: The school run also provides fertile ground for group action. You could start with your child's class, seeking out parents interested in car pooling or, better still, taking turns to lead a kids' crocodile walking to and from the school gates. Recruits would have to commit more time once or twice a week, but would no longer have to make two daily journeys. Plus the kids would get more exercise. Also try lobbying for sheltered and secure bicycle parking on school grounds so that older children can bike to lessons.

If the response is enthusiastic, you could organise a more formal school-based carbon diet club or ask the PTA to do so. This could involve parents, teachers and pupils in identifying other carbon-saving measures such as recycling paper, card and plastic supplies, printing double-sided on paper, planting trees and replacing old boilers.

In Your Community: Try setting up a diet club with people you see every week, perhaps neighbours or members of your book club, exercise class or local church. Just tap into the readymade networks in your community. Depending on how enthusiastic your recruits are, you could set very simple goals, such as one action a month by each member to cut both home energy and transport use. Or you could pass around a copy of this book and suggest that members work out their carbon weight, start following our standard diet and monitor their carbon weight loss.

FYI:

The numbers of staff driving to work at Orange in Bristol fell from 79% to 27% after the company began offering a car share matching service, cycling facilities and compensation for giving up a parking space.

Diet Pioneers

Get inspiration from these trailblazers...

Carbon Rationers: Their name may be a mouthful, but Carbon Rationing Action Groups (CRAGS) operate on a simple premise. Members of each local group agree to cut their CO_2 total by a specific amount a year (usually 10%). If they fail, they pay a small fine or other penalty. Personal emissions are calculated for home energy and transport use only, with a designated "carbon accountant" totting up gas and electricity bills and car and flight mileage. As of January 2006, CRAGs were operating in Oxford, Hereford, Leamington, Leeds, Brighton, Stroud, the Cotswolds, Worcestershire, Glasgow, Sevenoaks, Wokingham and North, West and South London.

Andy Ross, former carbon accountant for the Leamington CRAG, successfully cut his carbon calories by 10% in 2006. "I'm not a long in the tooth environmentalist," he emphasises. "I'm a civil engineer who has worked on motorway widening projects and got involved after reading a couple of books about climate change. Our group has achieved quite big cuts in our carbon use by making lifestyle changes that are significant, but not uncomfortable. For example, I drive much less now and travelled overland to Greece last summer for a wedding, rather than flying."

Interested in setting up a CRAG? Check out the network's ground rules, tips and upcoming events at www.carbonrationing.org.uk

Village Dieters: Ashton Hayes in rural Cheshire is seeking the title of Britain's first "carbon neutral" village. With help from the University of Chester residents calculated the village's total carbon weight at 4765.76 tonnes of CO_2 a year. It has since embarked on various communal actions including eco driving training days, tree planting schemes and home energy audits. After receiving national publicity the village is inspiring copycat action around the UK. Get inspired yourself by browsing at www.goingcarbonneutral.co.uk/.

Join an Online Diet Community

With climate change dominating the headlines, a variety of online pledge sites have sprung up. Here are a few of the best known. You might like to sign up with one or two as an extra incentive, while following our diet.

www.cred-uk.org CRed, the Community Carbon Reduction Project, is run by the University of East Anglia's School of Environmental Sciences. Register online to make pledges and receive tips and feedback on savings.

http://www.eclipse.co.uk/exeter/foe/eca_com.htm Run by Exeter Climate Action, this site features 35 commitments to pick and choose from, ranging from turning down your thermostat to buying a wind-up radio. A week after you sign up, your name will appear on the site's "Hall of Fame."

www.flightpledge.org.uk Make a gold or silver pledge and receive a certificate to print out. Gold requires you to take no flights (barring emergencies) in the 12 months following your pledge date; silver allows you two short haul or one long haul flight.

My World, accessible through the UK Sustainable Development Commission website at **www.sd-commission.org.uk/** Members record their actions to cut personal emissions on the website and receive feedback. Set up by the SDC, Eden Project and University of Surrey.

http://www.rsacarbonlimited.org/default.aspa Input your home and travel emissions and the Royal Society of Arts' Carbon Limited project will calculate your annual carbon weight. You can join in blogs, vote for your favourite calorie cutting ideas and amend your personal data as your diet progresses.

Demand Change

Pioneering groups of people can only achieve so much, however committed. To prevent runaway climate change we need government as well as citizen action. In 2006, Environment Secretary David Miliband floated the idea of carbon "credit cards" to encourage less petrol use, while Gordon Brown pledged that within 10 years all new homes would be "zero carbon." Yet the government has admitted it will miss a manifesto pledge to cut national CO_2 emissions by a fifth from 1990 levels by 2010.

So, carbon dieters, let's hold those politicians to their promises. When you have a spare few minutes, fill out the postcard on this page and send it off to your local councillor and MP. Encourage your friends to do the same.

The Low Carbon Diet

Dear ... ...
(name of councillor or MP)

To win my vote, please help your constituents
adopt low carbon diets:

• Introduce or improve home energy efficiency grants
• Improve and promote recycling facilities
• Improve public transport and cycle routes
• Vote for big tax breaks for hybrid cars
  and rooftop wind and solar energy
• Vote for airline fuel taxes

Yours, ... (Name)

(Address)

The Carbon Dieter's Directory

Introduction

Climate Change and Carbon Footprints

Carbon Footprint Ltd
at www.carbonfootprint.com
Learn about climate change, calculate your foot-print, offset your emissions.

Environmental Change Institute
at www.eci.ox.ac.uk/research/energy/
Energy research projects include renewable energy, energy behaviour and personal carbon trading.

An Inconvenient Truth
at www.climatecrisis.net/takeaction/
Follow up on the Al Gore film – lots of hot tips on how to reduce your emissions.

Climate Information Outreach Network (COIN)
at www.coinet.org.uk/
Accessible information on the impacts and challenges of climate change; also offers practical steps (including case studies) to reduce personal impacts.

Department of Environment (DEFRA)
at www.defra.gov.uk/environment/
and www.defra.gov.uk/environment/statistics/
DEFRA Helpline: 08459 33 55 77 or email: helpline@defra.gsi.gov.uk
Comprehensive general information and statistics on the environment.

UK Sustainable Development Commission, 2006: 'I Will if You Will'.
at www.sdcommission.org.uk/publications.php?id=367
Describes how the UK can shift to more sustainable lifestyles.

Mayer Hillman and Tina Fawcett 'How We Can Save The Planet', Penguin, 2004.
Provides a very good introduction to climate change science, public policy and citizen action.

Green Futures
at www.greenfutures.org.uk/
Tel: 01223 564334
Bimonthly magazine featuring climate policy news and practical solutions for reducing personal and business carbon emissions.

The National Energy Foundation
at www.nef.org.uk/energyadvice/co2emissions.htm
Map of global CO2 emissions and general information on climate change.

Guardian online environment section
at www.guardian.co.uk/environment
Database of articles includes all aspects of climate change.

Home

General Advice and Tips

The Energy Saving Trust (EST)

at www.est.org.uk/myhome
Tel: Energy Advice Centres 0800 512 012
Advice on improving your carbon footprint
at home, grants available in your area and
recommended installers and suppliers of
energy efficient products.

Or The Energy Saving Trust in Scotland

at www.est.org.uk/schri
Tel: 0800 138 8858

Centre for Alternative Technology

at www.cat.org.uk/information/
Tel: 01654 705989
Practical information and advice on all aspects of
low carbon living plus visit their Green Shop site.

The National Energy Foundation (NEF)

at www.nef.org.uk/
Tel: 01908 665 555
Hot tips on saving energy, renewables and
green electricity.

Low Impact Living Initiative

at www.lowimpact.org/index.htm
Useful factsheets on low carbon living, including
organic gardening, renewable energy, rainwater
harvesting and biodiesel vehicle fuel.

www.theyellowhouse.org.uk/

Story of how a family turned their 1930's ex-
council house into their environmental dream
home. Full of ideas for greening your home space.

www.energyfirst.org/Runningcosts.htm

An easy way to calculate the running costs of
your appliances.

www.moreassociates.com/energy/

Check out the latest in 'smart' electricity and gas
meters.

Renewable Energy

The Department of Trade and Industry's Low Carbon Buildings Programme

at www.lowcarbonbuildings.org.uk
Tel: 0800 915 0990
Grants provided for domestic renewable
technologies such as solar panels, solar water
heaters, wind turbines and wood stoves.

Solar Century

at www.solarcentury.com
Tel: 020 7803 0100
Suppliers of domestic solar systems, with
a network of recommended installers fully trained
to install solar PV systems throughout the UK.

British Wind Energy Association

at www.bwea.com
Tel: 020 7689 1960
Information and advice about wind energy
installations.

Water Use

Thames Water

at http://waterwise.fortune-cookie.com/
domestic/
Advice and hot tips on saving water in your home
and garden.

Green Electricity

at www.uswitch.com/Energy/Green-Energy.html
Find the right green electricity tariff for you.

Or try:

www.greenelectricity.org/
For more green electricity deals.

Recycling and waste

Waste Watch

at www.wastewatch.org.uk/
Tel: 020 7549 0300
Tons of information provided by national charity
campaigning for waste reduction.

www.wasteonline.org.uk/
Wacky waste facts provided by Waste Watch.

www.recycle-more.co.uk/
Tel: 0845 0682572
Advice and information on all aspects of recycling
at home, school and work. Key in where you live
and they'll tell you where you can recycle.

Recycle Now!

at www.recyclenow.com
Helpline: 0845 331 331
Same services as Recycle More above.

Green Gadgets

Save-a-Flush

at www.save-a-flush.co.uk

or Water Hippos

at www.hippo-the-watersaver.co.uk/
Cheap device that lowers your toilet flush volume.

www.megamanuk.com
State of the art energy efficient lights that give
off a similar light to incandescent bulbs.

www.rippleproducts.com/shop/
productsearch.asp?
Buy your shower timers here.

www.ecokettle.com
Get your green kettle here.

Garden

Water conservation

at www.taptips.ie

Tips on conserving water in the garden.

English Nature

at www.english-nature.org.uk/pubs/
publication/PDF/Wildlifegardening2.pdf

General guide to wildlife-friendly gardening.

Sustainable wood products

The Forest Stewardship Council

at www.fsc-uk.org

Tel: 01686 413 916

Key in what you want to buy and they'll tell you
where to find FSC certified products.

Wormeries and composting

Wiggly Wigglers

at www.wigglywigglers.co.uk

Tel: 0800 216 990 or 01981 500391

Or try:
Original Organics

at www.originalorganics.co.uk

Tel: 01884 841515

Recycle Works

at www.recycleworks.co.uk

Tel: 01254 820 088

Supplies a wide range of compost bins (many
made from FSC timber), wormeries and shredders.

The Composting Association

at http://www.compost.org.uk

Tel: 0870 160 3270

Organic gardening and grow your own fruit and veg

Henry Doubleday Research Association

at www.gardenorganic.org.uk/

Tel: 0247 630 3517

Advice on organic gardening and composting.

at www.organicgarden.org.uk

The Royal Horticultural Society

at www.rhs.org.uk/vegetables/diary.asp
and at www.rhs.org.uk/advice/watering.asp

Provides a handy vegetable gardener's calendar
and advice on drought tolerant plants.

BBC Gardening website

at www.bbc.co.uk/gardening/htbg/

How to be a good gardener: Alan Titchmarsh
helps you negotiate this online guide to growing
your own fruit and vegetables.

Green garden gadgets

Garden lights and furniture

at www.greenfingers.com

Tel: 0845 345 0728

For solar powered garden lights

www.wilsonsyard.com

Tel: 0289 269 2304

Great furniture from reclaimed materials.

www.buildingdesign.co.uk/arch-1/
marmax-products/plastic-fencing.htm

Tel: 01207 283 442

Garden furniture and fencing from recycled plastic.

Transport

General advice

Transport 2000
at www.transport2000.org.uk/
Tel: 020 7613 0743
National campaign group for sustainable transport.
Information on public policy and advice for
individuals and businesses on how to reduce
car travel.

Environmental Transport Association
at www.eta.co.uk
Tel: 0845 389 1010
Green alternative to the AA and RAC. Offers
insurance and breakdown services and campaigns
for more sustainable transport systems.

Calculating travel distances

Driving distances
at www.viamichelin.com
Go to the driving directions page, where you can
calculate door to door distances in the UK.

Driving distances
also at
www.theaa.com/travelwatch/planner_main.jsp

International distances
at www.indo.com/distance/index.html
Calculates international distances for flight and
other travel.

Walking distances in London
at www.walkit.com

Your car

Car CO2 emissions
at www.vcacarfueldata.org.uk
Find out your car's CO2 emissions and what
vehicle excise duty you'll have to pay.

www.yesinsurance.co.uk
Offers £25 cashback and carbon offsets to
owners of hybrid vehicles who take their insurance.

www.green-car-guide.com
Tips on greener driving plus the latest green
car models.

www.est.org.uk/fleet/Vehicles/Alternativefuels/
Alternativefuelsrefuellingmap/
Gives your nearest LPG (Liquified Petroleum Gas)
filling station.

www.goinggreen.co.uk
Listings site for green cars.

The AA
at www.theaa.com/allaboutcars/advice/advice_
rcosts_petrol_table.jsp
Find out what your car is costing you to run.

Global Action Plan
at www.globalactionplan.org.uk/
Click on 'Your Environment', then click on
'Transport'
Tel: 020 7405 5633
Smart driving tips and general environmental
advice.

www.whatcar.co.uk
Click on 'Help and Advice', then on 'Going Green'
Tips on buying a greener car.

Cooperative Insurance
at www.storedirect.co.uk/review/
insurance/cooperative/
Tel: 0845 600 4101 for a quote.
The Co-operative is offering the UK's first
eco-friendly car insurance. Offset 20% of your
emissions.

Carbon offsetting

Carbon offsetting schemes
at www.targetneutral.com
Pay through this BP scheme to cancel out your
emissions from car travel.

Car sharing clubs

www.nationalcarshare.co.uk.
Tel: 0871 8718 880

www.streetcar.co.uk

www.citycarclub.co.uk
Tel: 0845 3301 234

www.whizzgo.co.uk
Tel: 0870 446 6000

www.carplus.org.uk
Tel: 0113 234 9299
Smaller not for profit clubs listed on this site

Consumer

Food

The National Consumer Council
at www.ncc.org.uk/responsibleconsumption/
greening-supermarkets.pdf
Tel: 020 7730 3469
Report on greening supermarkets: how
supermarkets can make greener shopping easier.

Food Climate Research Network
at www.fcrn.org.uk/
Tel: 020 7686 2687
Researches and promotes ways to reduce
greenhouse gases across the UK food chain.

Soil Association
at www.soilassociation.org.uk/
Tel: 0117 314 5000
Organics facts and figures and a nationwide
directory of local food schemes.

Hugh Fearnley-Whittingstall
at www.rivercottage.net
Cooking with seasonal ingredients, box delivery
schemes and other delights.

**Farmers markets and fresh local
produce**
at www.farmersmarkets.net
Tel: 0845 4588 420
This site gives you your nearest certified farmer's
market

Or try **www.bigbarn.co.uk**
For local farmer's markets, delivery box schemes
and organic food shops.

www.farmshopping.com
Find your nearest farm selling fresh produce as well as box delivery schemes.

Shopping

Centre for Alternative Technology
at www.cat.org.uk/shopping
Tel: 01654 705 959
Extensive online catalogue for energy efficient home, garden and lifestyle products.

www.gooshing.co.uk
Compares thousands of products based on environmental criteria.

The Good Shopping Guide 2006
at www.thegoodshoppingguide.co.uk
Published by the Ethical Marketing Group, the Good Shopping Guide ranks the ethical scores of companies behind hundreds of the world's biggest brands. Anything from hoovers to hybrids, bread to boilers.

www.greenguideonline.com
Search online for green products.

www.edun.ie
Eco clothing company run by Bono's wife and stocked by Top Shop.

www.junkystyling.co.uk
Tel: 020 7247 1883
Gives worn or badly fitting clothes a contemporary makeover.

www.keepandshare.co.uk
Offers new clothes knitted from old.

Gifts

By Nature
at www.bynature.co.uk
Tel: 0845 456 7689
Online retailer of original eco-friendly and carbon-lite gifts.

www.goodgifts.org
Helpline: 020 7794 8000
All sorts of alternative gifts you can give, from regenerating English woodlands to providing school dinners for a class of African school children.

Carbon offsetting

www.climatecare.org.uk
(see listing under Holidays)

Pure – the Clean Planet Trust
at www.puretrust.org.uk
Helps individuals offset carbon emissions through the EU Emissions Trading Scheme.

Reuse and recycle

Computer Aid
at www.computer-aid.org
Tel: 020 7281 0091
Provides professionally refurbished computers for reuse in education, health and not-for-profit organisations in developing countries. Also receives payments from recyclers for donated mobile phones and used printer cartridges.

Action Aid Recycling

at www.actionaidreycling.org.uk/

Tel: 0117 304 2390

Action Aid Recycling collects ink and toner cartridges and mobile phones to raise funds for the charity Action Aid.

Furniture Reuse Network

at www.frn.org.uk/

Tel: 0117 954 3571

Coordinates a nationwide network of recycling groups which pass on unwanted furniture and electrical goods to low-income families.

Battery recycling

at www.rebat.com

Online nationwide register of recycling collection sites for nickel cadmium batteries maintained by the British Battery Manufacturers Association.

Green services

Norwich and Peterborough Building Society

at www.npbs.co.uk/about-us/ environmental-policy.asp

Offers a carbon neutral mortgage.

The Co-Op Bank

Call 0800 0288 288 for eco-mortgage details. Through ClimateCare, the Co-op will offset 20% of your home's energy use for the life of your mortgage.

www.greentomatocars.com

Tel: 020 8600 2520 for general information and 020 8748 8881 for bookings.

offers a London minicab service using hybrid cars.

Holidays

The Man in Seat Sixty One

at www.seat61.com

Website detailing rail links across and beyond Europe. Also details rail and ship links.

National Express coaches

at www.nationalexpress.com/home/hp.cfm

Tel: 08705 808 080

For details of travel by coach across Europe

Eurostar

at www.eurostar.com

www.sustrans.org.uk

Tel: 0845 113 0065

Coordinates National Cycling Network; provides detailed maps of scenic cycle routes.

www.responsibletravel.com

Tel: 01273 600 030

A large selection of eco holidays including walking holidays in the UK and in Europe.

www.exodus.co.uk

Tel: 0870 950 0039

Environmentally responsible travel company offering active holidays, including horse riding and mountain biking in Europe.

Carbon offsetting

British Airways' scheme (with Climate Care)

at http://www.ba.com/offsetyouremissions

Climate Care

at www.climatecare.org.uk

Group Dieting

Car sharing schemes

www.liftshare.org

www.shareajourney.com

www.energy-efficiency.gov.uk/transport
Tel: 0800 585 794

Setting up a carbon reduction action group

www.carbonrationing.org.uk
Find out more about carbon reduction action groups

www.goingcarbonneutral.co.uk/
Find out how the village of Ashton Hayes in Cheshire committed to going carbon neutral.

Online carbon diet clubs

www.cred-uk.org
The Community Carbon Reduction Project, run by University of East Anglia's School of Environmental Sciences. Commit to cutting your CO2 emissions by 60% by 2025.

www.eclipse.co.uk/exeter/foe/eca_com.htm
Run by Exeter Climate Action. Make a carbon reduction commitment here.

www.flightpledge.org.uk
Make a pledge to fly less here.

www.rsacarbonlimited.org/default.aspa
This site will calculate your carbon weight if you input home and travel emissions. Then you can vote for your favourite carbon calories cutting ideas.

Diet Masterplan Assumptions

(How we worked out the CO2 savings in the Diet Masterplan on page 135 and where we got the information to do our sums.)

Home Diet

Fit one energy saving light bulb
Source: The Energy Saving Trust at www.est.org.uk/myhome/whatcan/yourhome/.

Switch off idle electronics and appliances left on standby
Using Energy Saving Trust cost saving figures (£37/year). Assuming an electricity price of 10.41pence/kwh and using the DEFRA CO2 emissions factor for electricity (www.defra.gov.uk/environment/business/envrp/gas/envrpgas-annexes.pdf).

Unplug idle computers at night and w/ends
Using Energy Saving Trust cost saving figures (£35/year). Assuming an electricity price of 10.41pence/kwh and using DEFRA CO2 emissions factor for electricity.

Don't overfill the kettle. Boil what you need
A 2.5kw kettle taking 2.5 minutes to boil when full and 45 seconds to boil with 1cup water in. Kettle is boiled 5 times/day, 7 days/week. Using DEFRA CO2 emissions factor for electricity.

Turn down your washing machine to 40C rather than 60C/90C
Using National Energy Foundation figures (www.nef.org.uk/energyadvice/washing.htm) for kwh/wash for a B rated washing machine at 90C, 60C and 40C. Assuming 4 washes (or cycles) per week, 208 cycles/year and DEFRA CO2 emissions factor for electricity.

Halve the number of washes you put on and lower your wash temperature
Using Environmental Change Institute figures for Kwh/cycle for an average washing machine (www.eci.ox.ac.uk/research/energy/downloads/lowercarbonfuturereport.pdf

page 24, Table 2.8). Assuming 4 cycles/week, 208 cycles/year and 2 cycles/week, 104 cycles/year and using DEFRA CO2 emissions factor for electricity. And assuming going from a 90C to a 40C wash.

Run dishwasher on an Economy setting and halve the number of times you put it on

Using Environmental Change Institute figures for Kwh/cycle for an average dishwasher (www.eci.ox.ac.uk/research/energy/downloads/lowercarbonfuturereport.pdf page 24, Table 2.8). Assuming 260 cycles/year and 130 cycles/year and using DEFRA CO2 emissions factor for electricity.

Forget the tumble dryer, rediscover the clothesline

Using Environmental Change Institute figures for Kwh/cycle for an average tumble drier (www.eci.ox.ac.uk/research/energy/downloads/lowercarbonfuturereport.pdf page 24, Table 2.8). Assuming 208 cycles/year and using DEFRA CO2 emissions factor for electricity.

Replace old fridge freezer with an A+ or A+++ rated model

Source: the Energy Saving Trust, www.est.org.uk/myhome/whatcan/yourhome/

Fit a low flow shower head (family of four)

www.carbonbalanced.org/personal/pcbreduce.htm.

Turn hot water down to 60°C

Using Energy Saving Trust cost saving figures (£20/year). Assuming a gas price of 2.617pence/kwh and using DEFRA CO2 emissions factor for gas.

Turn heating down 1°C

Source: the Energy Saving Trust, www.est.org.uk/myhome/whatcan/yourhome/

Treat your hot water tank to an insulating jacket

Source: the Energy Saving Trust, www.est.org.uk/myhome/whatcan/yourhome/

Install a save-a-flush in your loo
Greenhouse gas emissions per litre of water treated and supplied to customers: 0.00029kg CO_2 equivalent (Thames Water). Installing a save-a-flush saves 2000 litres of water a year.

Spring cleaning ideas:

Draught-proof windows and doors
Source: the Energy Saving Trust, www.est.org.uk/myhome/whatcan/yourhome/

Seal holes in floors and skirting boards
Source: the Energy Saving Trust, www.est.org.uk/myhome/whatcan/yourhome/

Put foil behind your radiators
Assuming a gas price of 2.617pence/kwh and using DEFRA CO_2 emissions factor for gas.

Insulate your loft 10 inches deep
Source: the Energy Saving Trust, www.est.org.uk/myhome/whatcan/yourhome/

Install double glazing
Source: the Energy Saving Trust www.est.org.uk/myhome/whatcan/yourhome/

Upgrade to a condensing boiler
Source: the Energy Saving Trust www.est.org.uk/myhome/whatcan/yourhome/

Garden Diet

Swap the garden hose for a water butt

Greenhouse gas emissions per litre of water treated and supplied to customers: 0.00029kg CO_2 equivalent (Thames Water). Installing a water butt saves on average 1934.5 litres per year.

Compost food and garden waste

For every kilogram of waste you throw out, you produce 1 kg of CO_2. An average household throwing out 1 dustbin's worth of waste every week emits 1400kg of CO_2 a year. You can cut this figure by 20% if you compost all kitchen and garden waste. Source: Quaker Green Action, 2006.

Transport Diet

Ditch the car for journeys of 2km or less - walk or cycle (five times a week)

CO_2 saved as a result of not going on a 2km journey in an averaged sized petrol car 5 times/week, 20 times/month, 240 times/year (based on DEFRA CO_2 emissions factor for an averaged sized petrol car and assuming the journey is made during off-peak hours).

Take the train/bus/tube and not the car (12km/8m commute)

Difference in CO_2 emissions between taking an averaged sized petrol car on a 6km/4m (12km/8m roundtrip) commute to work during rush hour and doing the same journey by bus, diesel train and tube (the CO_2 emissions factors used are for commuting by car, bus, train and tube during peak hours and assume high occupancy in buses and trains. Figures are taken from Potter, Stephen (2004): Transport Energy and Emissions: Urban Public Transport, Chapter 13, pp 247-262 (Table 5) of Hensher, David and Button, Kenneth (Eds) Handbook of Transport and the Environment, Volume 4, Pergamon/Elsevier. Commute is made 5 times/week, 20 times/month and 210 times/year (assuming 6 weeks off per year).

Cycle to work

CO2 emissions saved by swopping a 6km/4m (12km/8m roundtrip) commute to work in an average sized petrol car during rush hour for the same journey by bike (the CO2 emissions factor used are for commuting by car during peak hours. Taken from Potter, Stephen (2004). Commute is made 5 times/week, 20 times/month and 210 times/year (assuming 6 weeks off per year).

Re-think the school run: walk/bike rather than drive (4km trip, twice a day)

CO2 emissions saved by biking/walking rather than taking an average sized petrol car on a 2km (4km roundtrip twice a day = 8km) journey to school 190 times a year (10 weeks off for school holidays). The CO2 emissions factor used are for commuting by car during peak hours. Taken from Potter, Stephen (2004).

Or...walk/bike 3 times a week

CO2 emissions saved by only driving to school (assumptions as above) twice a week. The CO2 emissions factor used are for commuting by car during peak hours. Taken from Potter, Stephen (2004).

Be a smooth driver: avoid sharp braking or acceleration

Source: Warwick University Climate Footprint Project at
http://www.carboncalculator.co.uk/reductions_transport.php
Explanation: Smooth driving can save 30% on fuel consumption, reducing carbon emissions – this means reducing the amount of sharp braking and sharp accelerating while you drive.

Consumer Diet

Most rubbish comes from food and other packaging. Recycle glass, paper, cardboard, plastics, cans.

For every kilogram of waste you throw out, you produce 1 kg of CO_2. An average household throwing out 1 dustbin's worth of waste every week emits 1400kg of CO_2 a year. You can cut this figure by 30% if you recycle all paper, glass, metal and plastic (apart from plastic bags). Source: Quaker Green Action, 2006.

Recycle your daily paper

Source: Warwick University's www.carboncalculator.co.uk

Do one weekly supermarket shop, not three

A 1.5km (3km roundtrip) journey to the supermarket in an averaged sized petrol car 3 times/week, 12 times/month and 144 times/year; and the same journey done once a week, 4 times/month and 36 times/year. Difference in CO_2 emissions between the 2 journey types using DEFRA CO_2 emissions factor for an average petrol car and assuming the journey is made during off-peak hours.

Buy British – a basket of UK grown food rather than produce flown from abroad

In British-grown basket: cauliflower from Lincolnshire, mushrooms from Ireland, brussel sprouts from Lincolnshire, broccoli from Worcestershire, carrots from Scotland and onions from Shropshire. In foreign-grown basket: limes from Brazil, pears from Italy, avocados from Chile, peaches from USA, pineapple from Costa Rica, baby corn from Kenya. Using DEFRA CO_2 emissions factor for long-haul air freight multiplied by a factor of 2.7 (to reflect the warming effect equivalent of other greenhouse gases in the upper atmosphere) and DEFRA CO_2 emissions factor for a 75% loaded articulated lorry. Assume an average lorry load of 9.3 tonnes.

Buy a kilo (2.2lb punnet) of British strawberries rather than Californian
Using DEFRA CO_2 emissions factor for long-haul air freight multiplied by a factor of 3 and DEFRA CO_2 emissions factor for a 75% loaded articulated lorry. Assume an average lorry load of 9.3 tonnes. Calculated using distance between San Jose, California and London (8657km) and between Canterbury and London (99km).

Buy a kilo of British pears, instead of South African grapes
Using DEFRA CO_2 emissions factor for long-haul air freight multiplied by a factor of 2.7 and DEFRA CO_2 emissions factor for a 75% loaded articulated lorry. Assume an average lorry load of 9.3 tonnes. Calculated using distance between Johannesburg and London (9027km) and between Canterbury and London (99km).

Buy a kilo of British green beans, not Kenyan
Using DEFRA CO_2 emissions factor for long-haul air freight multiplied by a factor of 2.7 and DEFRA CO_2 emissions factor for a 75% loaded articulated lorry. Assume an average lorry load of 9.3 tonnes. Calculated using distance between Nairobi and London (6804km) and between Canterbury and London (99km)

Buy a bottle of French wine instead of a New Zealand vintage
Using DEFRA CO_2 emissions factor for large bulk carrier (ship) freight and DEFRA CO_2 emissions factor for a 75% loaded articulated lorry. Assume an average lorry load of 9.3 tonnes. Calculated using distance between Auckland and London (18,331km) and between Bordeaux and London (742km).

Holiday Diet

Take the train for a UK or short overseas trip (350km x 2), not the plane
CO_2 saved as a result of taking a journey of 700km roundtrip by train rather than by plane, using DEFRA CO_2 emissions factor for train and the DEFRA CO_2 figure for plane travel multiplied by a factor of 2.7 to reflect the warming effect equivalent of other greenhouse gases in the upper atmosphere.

Or...take a coach, not the plane
CO_2 saved as a result of taking a journey of 700km roundtrip by coach rather than by plane, using DEFRA CO_2 emissions factor multiplied by a factor of 2.7 for plane travel; and DfT CO_2 emissions for a coach.

Take Eurostar, not the plane, from London to Paris or Brussels
From Eurostar press release at:
http://www.eurostar.com/UK/uk/leisure/about_eurostar/press_release/
02_10_06_environment.jsp
Eurostar figures quoted on Seat 61 website at www.seat61.com/

Take the train rather than fly London to Edinburgh
Using DEFRA CO_2 emissions factor for train and the DEFRA CO_2 figure for plane travel multiplied by a factor of 2.7.

Take 1 rather than 3 short haul (500km) flights a year
CO_2 saved as a result of taking 1 trip of 1000km (500km each way) instead of 3 trips of 1000km (500km each way). Using DEFRA CO_2 emissions factor for air travel multiplied by a factor of 2.7.

Advanced Diet – Carbon Supersavers

Install cavity wall insulation
Using Energy Saving Trust cost saving figures (£160/year). Assuming a gas price of 2.617pence/kwh and using DEFRA CO2 emissions factor for gas.

Switch to a green electricity supplier
Source: Warwich University Climate Footprint Project, www.carboncalculator.co.uk

Install rooftop wind turbine for DIY electricity
Source: Faber Maunsell quoted in the Guardian 21/10/06 'Hot air and the dash from gas'

Install rooftop solar panels for DIY electricity
Source: Faber Maunsell quoted in the Guardian 21/10/06 'Hot air and the dash from gas'

Buy a hybrid car
Difference between an average petrol car emitting 180gCO2/km (DEFRA) and a Honda Civic emitting 116gCO2/km (Ethical Consumer Guide).

Sell your car, use public transport
Using DEFRA CO2 emissions factor for an average petrol car and an average CO2 emissions factor using a mix of urban diesel train, urban electric train and bus (using Potter et al CO2 emissions factors). Difference in CO2 emitted between travel in a car over 16,000km (10,000m) and travel on public transport over 16,000km (10,000m).

Give up beef
Beef is particularly bad because of cows' digestion which produces methane – a very powerful greenhouse gas. For each 0.45kg (1lb) of beef, a cow has produced methane with the equivalent warming effect of 5.2kg CO2. Source: www.coinet.org.uk

Well done

Polly Ghazi is the former Environment Correspondent
of The Observer and co-author of the bestselling *Downshifting:
The Guide to Happier, Simpler Living* (Hodder & Stoughton 1997).
She currently lives in Washington.

Rachel Lewis is an environmental consultant who has worked
for the World Bank, the CBI and a grassroots environmental charity.
She lives in West London.

**In case of difficulty in purchasing any Short Books
title through normal channels, please contact
BOOKPOST**

Tel: 01624 836000

Fax: 01624 837033

email: bookshop@enterprise.net

www.bookpost.co.uk

Please quote ref. 'Short Books'